SpringerBriefs in Electrical and Computer Engineering

T0213813

More information about this series at http://www.springer.com/series/10059

Yi Song • Jiang Xie

Broadcast Design in Cognitive Radio Ad Hoc Networks

 Springer

Yi Song
Wichita State University
Wichita, Kansas
USA

Jiang Xie
The University of North Carolina at Charlotte
Charlotte, North Carolina
USA

ISSN 2191-8112   ISSN 2191-8120 (electronic)
SpringerBriefs in Electrical and Computer Engineering
ISBN 978-3-319-12621-0   ISBN 978-3-319-12622-7 (eBook)
DOI 10.1007/978-3-319-12622-7

Library of Congress Control Number: 2014948601

Springer Cham Heidelberg New York Dordrecht London

Printed on acid-free paper

Springer is part of Springer Science+Business Media (www.springer.com)

Preface

Broadcast is an important operation in wireless networks where control information is usually propagated as broadcasts for the realization of most networking protocols. In traditional ad hoc networks, since the spectrum availability is uniform, broadcasts are delivered via a common channel which can be heard by all users in a network. However, in cognitive radio (CR) ad hoc networks, unlicensed users may observe heterogeneous spectrum availability, which is unknown to other unlicensed users before the control information was broadcast. This non-uniform spectrum availability imposes special design challenges for broadcast in CR ad hoc networks. In this book, the special challenges of broadcast design in CR ad hoc networks are discussed. Two broadcast protocols in CR ad hoc networks are introduced: (1) a quality-of-service (QoS)-based broadcast protocol under blind information and (2) a fully-distributed broadcast protocol with collision avoidance. In addition, a novel unified analytical model is also introduced to analyze the performance of broadcast protocols in CR ad hoc networks. This is the first book that discusses the unique broadcast challenges in CR ad hoc networks and investigates the analytical work on the performance analysis of broadcast protocols for CR ad hoc networks.

Contents

1 Introduction 1
  1.1 Background on Cognitive Radio Networks 1
  1.2 Background on Broadcast in Computer Networks 2
      1.2.1 Existing Broadcast Protocols in Traditional Wireless
            Networks 2
  1.3 Broadcast Protocols in Cognitive Radio Networks 3
      1.3.1 Research Status 3
      1.3.2 Research Challenges 3
  1.4 Analytical Model for Broadcast Protocols in Cognitive Radio
      Networks 5
      1.4.1 Research Status 5
      1.4.2 Research Challenges 6
  1.5 Contributions 8
  1.6 Organization 9
  References 9

2 QoS-based Broadcast Protocol Under Blind Information in Cognitive
  Radio Ad Hoc Networks 13
  2.1 Network Model 13
  2.2 Exploring Broadcast Design in CRNs 14
      2.2.1 Random Broadcast Scheme 14
      2.2.2 Full Broadcast Scheme 16
  2.3 The Basic Scheme 17
      2.3.1 The Single-Hop Scenario 17
      2.3.2 The Multi-Hop Scenario 20
  2.4 The Enhanced Scheme 21
      2.4.1 Analysis of the Channel Availability 21
      2.4.2 The Enhanced QB^2IC Scheme 25
  2.5 Performance Evaluation 25
  References 36

**3 Distributed Broadcast Protocol with Collision Avoidance in Cognitive
    Radio Ad Hoc Networks** 37
    3.1 The Distributed Broadcast Protocol with Collision Avoidance 37
         3.1.1 Construction of the Broadcasting Sequences 37
         3.1.2 The Distributed Broadcast Scheduling Scheme 41
         3.1.3 The Broadcast Collision Avoidance Scheme 42
    3.2 The Derivation of the Value of w 46
         3.2.1 The Network Model 46
         3.2.2 The Derivation of the Value of w 47
    3.3 Discussion on the Proposed Broadcast Protocol 52
         3.3.1 The 2-hop Location Information 52
         3.3.2 Time Synchronization 53
    3.4 Performance Evaluation 55
    References 64

**4 Unified Analytical Model for Broadcast in Cognitive Radio Ad Hoc
    Networks** 67
    4.1 Calculating the Successful Broadcast Ratio 67
         4.1.1 The Unique Challenge 67
         4.1.2 The Proposed Algorithm 69
         4.1.3 An Illustrative Example 70
    4.2 Calculating the Average Broadcast Delay 72
         4.2.1 The Unique Challenge 72
         4.2.2 The Proposed Algorithm 73
         4.2.3 An Illustrative Example 76
    4.3 Broadcasting in CR Ad Hoc Networks 76
         4.3.1 Random Broadcast Scheme 76
         4.3.2 QoS-based Broadcast Scheme 78
         4.3.3 Distributed Broadcast Scheme 81
    4.4 Performance Evaluation 83
         4.4.1 Validating Analysis Using Hardware Implementation 83
         4.4.2 Validating Analysis Using Simulation 85
         4.4.3 System Parameter Design Using the Proposed Analytical
                Model 90
    References 93

5 Conclusion 95

Chapter 1
Introduction

1.1 Background on Cognitive Radio Networks

The invention of portable hand-held devices, such as smartphones and tablet personal computers (PCs), has led a dramatic increase in the demand of ubiquitous wireless communication networks and has revolutionized the way people communicate these days. Ranging from old-fashioned cellular networks (e.g., 4G networks) to recent local area networks (e.g., WiFi networks) or body area networks (e.g., Bluetooth), various types of wireless network services are seeing an unprecedented growth. With such rapid growth of wireless devices, the demands for the radio spectrum are constantly increasing, resulting in scarce spectrum resources. According to the Federal Communications Commission (FCC), almost all the radio spectrum for wireless communications has already been allocated. However, recent studies show that up to 85 % of the allocated spectrum is underutilized due to the current fixed spectrum access policy [1]. To alleviate the spectrum scarcity problem, FCC has suggested a new paradigm for dynamically accessing the vacant portions of the allocated spectrum [2]. Cognitive radio (CR) has recently emerged as a promising technology to overcome the imbalance between the increase in spectrum access demand and the inefficiency in spectrum usage by allowing dynamic spectrum access (DSA). A "cognitive radio" is a radio that can change its communication protocol parameters (e.g., operating frequency) based on interactions with the environment in which it operates [2–4]. CR networks are regarded as the next-generation wireless networks to efficiently utilize the radio spectrum. With the capability of sensing the frequency bands in a time and location-varying spectrum environment and adjusting the operating parameters based on the sensing outcome, CR technology allows an unlicensed user (or, secondary user) to exploit those frequency bands unused by licensed users (or, primary users) in an opportunistic manner [5]. Secondary users (SUs) can form a CR infrastructure-based network or a CR ad hoc network. Recently, CR ad hoc networks have attracted plentiful research attention due to their various applications [6, 7].

© The Author(s) 2014   1
Y. Song, J. Xie, *Broadcast Design in Cognitive Radio Ad Hoc Networks*,
SpringerBriefs in Electrical and Computer Engineering, DOI 10.1007/978-3-319-12622-7_1

1.2   Background on Broadcast in Computer Networks

Broadcast is an important operation in computer networks, especially in distributed multi-hop multi-channel wireless networks. Control information exchange among nodes, such as channel availability and routing information, is crucial for the realization of most networking protocols in an ad hoc network. This control information is often sent out as network-wide *broadcasts*, messages that are sent to all other nodes in a network. In addition, some exigent data packets such as emergency messages and alarm signals are also delivered as network-wide broadcasts [8].

1.2.1   Existing Broadcast Protocols in Traditional Wireless Networks

In traditional mobile ad hoc networks (MANETs), broadcast messages are usually conveyed through a common channel which can be heard by all nodes in the network [9–12]. Typical broadcast protocols in traditional wireless networks can be categorized into three categories. The first category is called *simple flooding*. That is, every node in the network rebroadcasts the broadcast message with a probability of one when it receives the message. This method leads to a very high overhead. More severely, sometimes it leads to a failed broadcast. That is, if a node receives two or more broadcast messages simultaneously, none of these messages can be successfully received by the node. This is called the *broadcast collision problem*. The second category is called the *probability-based method*. In this method, each node rebroadcasts the message with a probability of *p* when it receives the message. It can reduce the overall overhead and the probability of broadcast collisions. In dense networks, multiple nodes share similar transmission coverages. Thus, the probability-based method can reserve network resource without harming delivery effectiveness. However, in sparse networks, the broadcast delivery effectiveness may be affected. The third category is called the *neighbor knowledge method*. In this method, some network topology information is given, such as the locations of 1-hop nodes, 2-hop nodes or sometimes the whole network topology. Based on this information, some nodes can be particularly selected as the forwarding nodes while other nodes do not rebroadcast the broadcast message. These forwarding nodes often can lead to full coverage and avoid broadcast collisions. In addition, it can also lead to the shortest broadcast delay. However, a-priori network topology information is required.

Since broadcast messages often need to be disseminated to all destinations as quickly as possible, we aim to achieve very high successful broadcast ratio (i.e., the probability that all nodes in a network successfully receive a broadcast message) and very short average broadcast delay (i.e., the average duration from the moment a broadcast starts to the moment the last node receives the broadcast message).

1.3 Broadcast Protocols in Cognitive Radio Networks

1.3.1 Research Status

Currently, research on broadcast protocols in multi-hop CR ad hoc networks is still in its infant stage. There are only limited papers addressing the broadcast issue in CR ad hoc networks [13–16]. However, in [13] and [14], the global network topology and the available channel information of all SUs are assumed to be known. Additionally, in [14], a common signaling channel for the whole network is employed which is also not practical. These two papers adopt impractical assumptions which make them inadequate to be used in practical scenarios. Other proposals aiming to locally establish a common control channel may also be considered for broadcast [17–20]. However, these proposals need a-priori channel availability information of all SUs which is usually obtained via broadcasts. In addition, although some schemes on channel hopping in CR networks can be used for finding a common channel between two nodes [21–23], they still suffer various limitations and cannot be used in broadcast scenarios. In [21] and [22], the proposed channel hopping schemes cannot guarantee rendezvous under some special circumstances. In addition, one of the proposed schemes in [21] only works when two SUs have exactly the same available channel sets. Furthermore, in [23], a jump-stay based channel hopping algorithm is proposed for guaranteed rendezvous. However, the expected rendezvous time for the asymmetric model (i.e., different users have different available channels) is of polynomial complexity with respect to the total number of channels. Thus, it is unsuitable for broadcast scenarios in CR ad hoc networks where channel availability is usually non-uniform and short broadcast delay is often required. Other channel hopping algorithms explained in [24] require tight time synchronization which is also not feasible before any control information is exchanged.

1.3.2 Research Challenges

In CR ad hoc networks, since secondary users can only use the channels which are not occupied by primary users (PUs), different SUs may acquire different sets of available channels. Thus, the availability of such a common channel for all nodes may not exist. More importantly, before control information is sent, a SU is unaware of the available channels of its neighboring nodes. As a result, even though a global or local common channel may exist, SUs are unaware of its existence. Therefore, broadcasting control messages on a common channel is often not feasible in CR ad hoc networks under blind information [17, 18, 25–29].

On the other hand, since each SU is typically equipped with one radio, it can only visit one available channel at a time. A broadcast message can be received by a neighboring node only if the message is transmitted on the same channel on which the receiver currently stays. Therefore, if a SU only broadcasts a message once on one available channel, this message may not be successfully received by all its neighboring nodes. In addition, the number of times that a SU needs to broadcast the message is usually not deterministic. That is, the sender usually does not know the exact time when a broadcast to all its neighboring nodes is finished. This is because of the following two reasons. First of all, even though some feedback messages (e.g., acknowledgment messages (ACKs)) are used to inform the sender of successful broadcast deliveries in traditional ad hoc networks [30, 31], applying ACK messages in broadcast scenarios for CR ad hoc networks may not be feasible. If all receivers are required to send ACKs in response to the reception of a broadcast message, these ACKs may cause channel congestion and packet collision, which is known as the *ACK implosion problem* [32]. Secondly, in CR ad hoc networks, since different SUs may stay on different channels, before control information is sent, the SU sender does not have the channel information of its neighboring nodes. Therefore, the sender does not know which channel the receiver can use to receive the message. As a result, the sender does not have a determined order for every neighboring node to receive the message. Thus, the exact time for all neighboring nodes to successfully receive the broadcast message is also unknown. As a result, in CR ad hoc networks, the SU sender needs to broadcast the message multiple times to increase the probability that all its neighboring nodes successfully receive the messages.

In addition, in CR ad hoc networks, different SUs may acquire different sets of available channels. This non-uniform channel availability imposes special design challenges for broadcast in CR ad hoc networks. First of all, for traditional single-channel and multi-channel ad hoc networks, due to the uniformity of channel availability, all nodes can tune to the same channel. Thus, broadcast messages can be conveyed through a single common channel which can be heard by all nodes in a network. However, in CR ad hoc networks, the availability of a common channel for all nodes may not exist. More importantly, before any control information is exchanged, a SU is unaware of the available channels of its neighboring nodes. Therefore, broadcasting messages on a global common channel is not feasible in CR ad hoc networks.

To further illustrate the challenges of broadcast in CR ad hoc networks, we consider a single-hop scenario shown in Fig. 1.1, where node A is the source node. For traditional single-channel and multi-channel ad hoc networks, as shown in Fig. 1.1(a), nodes can tune to the same channel (e.g., channel 1) for broadcasting. Thus, node A only needs one time slot to let all its neighboring nodes receive the broadcast message in an error-free environment. However, in CR ad hoc networks where the channel availability is heterogeneous and SUs are unaware of the available channels of each other, as shown in Fig. 1.1(b), node A may have to use multiple channels for broadcasting and may not be able to finish the broadcast within one time slot. In fact, the exact broadcast delay for all single-hop neighboring nodes to successfully receive the broadcast message in CR ad hoc networks relies on various factors (e.g., channel availability and the number of neighboring nodes) and it is random.

Furthermore, in multi-hop scenarios, packet collisions occur when a node receives multiple broadcast messages simultaneously. This is called the *broadcast collision problem* [10]. Due to the absence of collision detection (CD) in wireless communications and network topology information (e.g., the identities of the neighboring nodes),

Fig. 1.1 The single-hop broadcast scenario

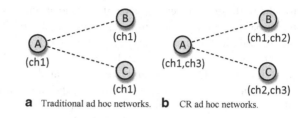

a Traditional ad hoc networks.   **b** CR ad hoc networks.

broadcast collisions are difficult to avoid. A broadcast protocol should be able to efficiently mitigate the broadcast collisions in multi-hop scenarios due to the waste of network resources and the performance degradation. Therefore, to design a broadcast protocol with satisfactory performance (e.g., high success rate and short broadcast delay, which indicates lower number of collisions) is critical for multi-hop CR ad hoc networks. Since multiple channels may be used for broadcasting and the exact time for all single-hop neighboring nodes to successfully receive the broadcast message is random, to avoid broadcast collisions (i.e., a node receives multiple copies of the broadcast message simultaneously) is much more complicated in CR ad hoc networks, as compared to traditional ad hoc networks. In traditional ad hoc networks, numerous broadcast scheduling schemes are proposed to reduce the probability of broadcast collisions while optimizing the network performance [26–28, 33–35]. All these proposals are on the basis that all nodes use a single channel for broadcasting and the exact delay for a single-hop broadcast is one time slot. However, in CR ad hoc networks, without the information about the channel used for broadcasting and the exact delay for a single-hop broadcast, to predict when and on which channel a broadcast collision occurs is extremely difficult. Hence, to design a broadcast protocol which can avoid broadcast collisions, as well as provide high successful broadcast ratio and short broadcast delay is a very challenging issue for multi-hop CR ad hoc networks under practical scenarios. Simply extending existing broadcast protocols to CR ad hoc networks cannot yield the optimal performance.

1.4 Analytical Model for Broadcast Protocols in Cognitive Radio Networks

1.4.1 Research Status

Even though the broadcast issue has been studied extensively in traditional mobile ad hoc networks (MANETs) [9–12, 36] these proposals mainly focus on broadcast protocol designs. The performance analysis of these proposed protocols is simulation-based. Thus, the analytical relationship between these proposals and their performance is not known. More importantly, without analytical analysis, the system parameters in these protocols are not designed to achieve the optimal performance. In fact, analytical analysis is beneficial not only for better understanding the nature of a proposed protocol, but also for better designing the system parameters of a protocol to achieve the optimal performance. It can also provide useful insights to guide the future broadcast protocol designs in CR ad hoc networks. Hence, in this book, we focus on the analytical analysis of broadcast protocols for multi-hop CR ad hoc networks.

Although a vast amount of analytical works on broadcast protocols in traditional MANETs exist [37–41], currently, there is no analytical work on broadcast protocols in multi-hop CR ad hoc networks. More importantly, all the methods proposed for traditional MANETs cannot be simply applied to multi-hop CR ad hoc networks. This is because that in traditional MANETs, the channel availability is uniform for all nodes, as shown in Fig. 1.1(a). However, in CR ad hoc networks, different SUs may acquire different available channel sets, depending on the locations and traffic of PUs, as shown in Fig. 1.1(b). This non-uniform channel availability leads to several significant differences and causes unique challenges when analyzing the performance of broadcast protocols in CR ad hoc networks.

Furthermore, no existing work on CR ad hoc networks addresses these challenges. Moreover, due to the above explained differences, the analytical methodology for broadcast protocol analysis in tradition MANETs cannot be extended to CR ad hoc networks. Specifically, the existing performance analytical papers on broadcasting in traditional multi-channel ad hoc networks cannot reflect the unique features (e.g., non-uniform channel availability and channel rendezvous schemes) in multi-hop CR ad hoc networks because: (1) a common control channel is used for broadcasting [42–46]; (2) only single-hop scenario is considered [42, 44, 47]; (3) a centralized entity is needed to schedule the broadcast [47]; and (4) multiple radios are used [48].

1.4.2 Research Challenges

First of all, unlike in traditional MANETs, in CR ad hoc networks, the single-hop broadcast is not always successful in an error-free environment. The reason can be illustrated using Fig. 1.1. If node A is the source node, in traditional MANETs, all its neighboring nodes can tune to the same channel to receive the broadcast message. However, in CR ad hoc networks, such a common available channel for all neighboring nodes may not exist [19–23]. As a result, the broadcast may fail. More severely, even if a common available channel exists between the source node and its neighboring nodes, they may not be able to tune to that channel at the same time, which will also result in a failed broadcast. In fact, whether the single-hop broadcast is successful depends on the channel availability of each SU which is time-varying and location-varying. Due to the uncertainty of the single-hop broadcast success, the successful broadcast ratio of a network is usually random. Furthermore, since there usually exist multiple message propagation scenarios for all the nodes to successfully receive the broadcast message in a multi-hop CR ad hoc network, it is extremely challenging to identify every possible message propagation scenario for calculating the successful broadcast ratio in a complicated network. An example illustrating this challenge will be given in Chap. 4.1.1.

Secondly, different from traditional MANETs where the relative locations of the communication pair do not impact the successful receipt of the message as long as they are within the transmission range of each other, in CR ad hoc networks, the probability that a node successfully receives a broadcast message is affected by the relative locations between the sender and the receiver. This is because that the available channels of a SU are obtained based on the sensing outcome from the proximity of the node. Thus, SU nodes that are close to each other have similar available channels and they may have higher successful broadcast ratio, as compared with the SU nodes far away from each other whose available channels are often less similar. These two differences show that the successful broadcast ratio is affected by various factors and it is random. Currently, there is no straightforward solution to analyze this issue.

Thirdly, the single-hop broadcast delay is usually more than one time slot in CR ad hoc networks, while in traditional MANETs, it is always one time slot. As shown in Fig. 1.1a, node A only needs one time slot to let all its neighboring nodes receive the broadcast message in an error-free environment. However, in CR ad hoc networks, due to the non-uniform channel availability, node A may have to use multiple channels for broadcasting and may not be able to finish the broadcast within one time slot. In fact, the exact broadcast delay for all single-hop neighboring nodes to successfully receive the broadcast message in CR ad hoc networks relies on various factors (e.g., channel availability and the number of neighboring nodes) and it is also random. Moreover, since there may exist multiple message propagation scenarios, to identify which node is the last node in a network to receive the message is very difficult. Thus, the multi-hop broadcast delay is extremely difficult to obtain.

Finally, broadcast collisions are complicated in CR ad hoc networks. Unlike in traditional MANETs where nodes use a common channel for broadcasting, in CR ad hoc networks, nodes may use multiple channels for broadcasting. Without the information about the channel used for broadcasting and the exact delay for a single-hop broadcast, to predict when and on which channel a broadcast collision occurs is extremely difficult. Hence, to mathematically analyze broadcast collisions is very challenging for multi-hop CR ad hoc networks under practical scenarios.

In summary, due to the randomness of the single-hop successful broadcast ratio and broadcast delay, the broadcast performance of a multi-hop CR ad hoc network is extremely challenging to analyze. Therefore, in this book, we study the performance analysis of broadcast protocols for multi-hop CR ad hoc networks. A novel unified analytical model is proposed to analyze the broadcast protocols in CR ad hoc networks with any topology. Specifically, in this book, we propose to decompose an intricate network into several simple networks which are tractable for analysis. We also propose systematic methodologies for such decomposition.

1.5 Contributions

This book studies the broadcast issues in cognitive radio ad hoc networks. The contributions of our work are as follows:

- A *Quality-of-Service (QoS)-based broadcast protocol under blind information* for multi-hop CR ad hoc networks, QB^2IC, is proposed. Two schemes of the QB^2IC protocol are introduced. In our design, we consider practical scenarios: (1) the network topology is not known; (2) the channel information of other SUs is not known; (3) the available channel sets of different SUs are not assumed to be the same; and (4) tight time synchronization is not required. In a word, blind information is considered for our design. In addition, by utilizing the diversity of channels, broadcast collisions can be mitigated in multi-hop scenarios under our proposed broadcast protocol. Furthermore, our proposed QB^2IC protocol can also work in a radio environment where transmission errors exist by adjusting its parameters.
- A *fully-distributed* broadcast protocol in a multi-hop CR ad hoc network, BRACER, is proposed. By intelligently designing the number of channels used for broadcasting, our proposed BRACER protocol can provide very high successful delivery ratio while achieving very short broadcast delay. In addition, a novel algorithm is proposed so that the broadcast collisions can be totally avoided.
- *An algorithm for calculating the successful broadcast ratio* (i.e., the probability that all nodes in a network successfully receive a broadcast message) is proposed for CR ad hoc networks. The proposed algorithm is a general methodology that can be applied to any broadcast protocol proposed for multi-hop CR ad hoc networks with any topology. In addition, *An algorithm for calculating the average broadcast delay* (i.e., the average duration from the moment a broadcast starts to the moment the last node in the network receives the broadcast message) is proposed for CR ad hoc networks under grid topology. The derivation methods of the single-hop performance metrics, successful broadcast ratio, average broadcast delay, and broadcast collision rate (i.e., the probability that a single-hop broadcast fails due to broadcast collisions), for three different broadcast protocols in CR ad hoc networks *under practical scenarios* (e.g., no dedicated common control channel exists and the channel information of any other SUs is not known) are proposed. Finally, *A hardware system is developed to implement different broadcast protocols* in multi-hop CR ad hoc networks and validate our proposed unified analytical model.

To the best of our knowledge, this is the first book that addresses the broadcast challenges specifically in multi-hop CR ad hoc networks and the first analytical work on the performance analysis of broadcast protocols in multi-hop CR ad hoc networks.

1.6 Organization

The rest of this book is organized as follows. A QoS-based broacast protocol under blind information in CR ad hoc networks is proposed in Chap. 2. In addition, A distributed broadcast protocol with collision avoidance in CR ad hoc networks is presented in Chap. 3. In Chap. 4 a unified analytical model for broadcast in CR ad hoc networks is introduced. Finally, we conclude this book in Chap. 5.

References

1. FCC, "Notice of proposed rule making and order," December 2003.
2. J. M. III "Cognitive radio: an integrated agent architecture for software defined radio," Ph.D. dissertation, KTH Royal Institute of Technology, Sweden, 2000.
3. J. M. III and G. Q. Maguire, "Cognitive radio: making software radios more personal," *IEEE Personal Communications*, pp. 13–18, August 1999.
4. S. Haykin "Cognitive radio: brain-empowered wireless communications," *IEEE Journal on Selected Areas in Communications (JSAC)*, vol. 23, no. 2, pp. 201–220, February 2005.
5. I. F. Akyildiz, W.-Y. Lee, M. C. Vuran, and S. Mohanty "NeXt generation/dynamic spectrum access/cognitive radio wireless networks: A survey," *Computer Networks (Elsevier)*, vol. 50, 2006.
6. I. F. Akyildiz, W.-Y. Lee, and K. R. Chowdhury "CRAHNs: cognitive radio ad hoc networks," *Ad Hoc Networks*, vol. 7, no. 5, pp. 810–836, July 2009.
7. Y. Song and J. Xie, *Cognitive Radio Mobile Ad Hoc Networks*. Springer, 2011, ch. On the Spectrum Handoff for Cognitive Radio Ad Hoc Networks without Common Control Channel.
8. G. Resta, P. Santi, and J. Simon "Analysis of multi-hop emergency message propagation in vehicular ad hoc networks," in *Proc.* ACM MobiHoc, 2007, pp. 140–149.
9. I. Chlamtac and S. Kutten, "On broadcasting in radio networks — problem analysis and protocol design," *IEEE Transactions on Communications*, vol. 33, no. 12, pp. 1240–1246, Dec. 1985.
10. S.-Y. Ni, Y.-C. Tseng, Y.-S. Chen, and J.-P. Sheu "The broadcast storm problem in a mobile ad hoc network," in *Proc.* ACM MobiCom, 1999, pp. 151–162.
11. J. Wu and F. Dai, "Broadcasting in ad hoc networks based on self-pruning," in *Proc.* IEEE INFOCOM, 2003, pp. 2240–2250.
12. J. Qadir, A. Misra, and C. T. Chou "Minimum latency broadcasting in multi-radio multi-channel multi-rate wireless meshes," in *Proc. IEEE SECON*, vol. 1, 2006, pp. 80–89.
13. Y. Kondareddy and P. Agrawal, "Selective broadcasting in multi-hop cognitive radio networks," in *Proc.* IEEE Sarnoff Symposium, 2008, pp. 1–5.
14. C. J. L. Arachchige, S. Venkatesan, R. Chandrasekaran, and N. Mittal "Minimal time broadcasting in cognitive radio networks," in *Proc. ICDCN*, 2011, pp. 364–375.
15. Y. Song and J. Xie, "A QoS-based broadcast protocol for multi-hop cognitive radio ad hoc networks under blind information," in *Proc.* IEEE GLOBECOM, 2011, pp. 1–5.
16. Y. Song and J. Xie, "QB^{2}IC: A QoS-based broadcast protocol under blind information for multi-hop cognitive radio ad hoc networks," *IEEE Transactions on Vehicular Technology*, vol. 63, no. 3, pp. 1453–1466, March 2014.
17. L. Lazos, S. Liu, and M. Krunz "Spectrum opportunity-based control channel assignment in cognitive radio networks," in *Proc.* IEEE SECON, 2009, pp. 1–9.
18. J. Zhao, H. Zheng, and G. Yang "Spectrum sharing through distributed coordination in dynamic spectrum access networks," *Wireless Communications and Mobile Computing*, vol. 7, pp. 1061–1075, November 2007.

19. K. Bian, J.-M. Park, and R. Chen "Control channel establishment in cognitive radio networks using channel hopping," *IEEE JSAC*, vol. 29, no. 4, pp. 689–703, April 2011.
20. Y. Zhang, Q. Li, G. Yu, and B. Wang "ETCH: Efficient channel hopping for communication rendezvous in dynamic spectrum access networks," in *Proc.* IEEE INFOCOM, 2011, pp. 2471–2479.
21. N. Theis, R. Thomas, and L. DaSilva "Rendezvous for cognitive radios," *IEEE Trans. Mobile Computing*, vol. 10, no. 2, pp. 216–227, 2010.
22. C. Cormio and K. R. Chowdhury, "Common control channel design for cognitive radio wireless ad hoc networks using adaptive frequency hopping," *Ad Hoc Networks*, vol. 8, no. 4, pp. 430–438, 2010.
23. Z. Lin, H. Liu, X. Chu, and Y.-W. Leung "Jump-stay based channel hopping algorithm with guaranteed rendezvous for cognitive radio networks," in *Proc.* IEEE INFOCOM, 2011.
24. J. Mo, H.-S. W. So, and J. Walrand "Comparison of multichannel MAC protocols," *IEEE Trans. on Mobile Computing*, vol. 7, no. 1, 2008.
25. H.-P. Shiang and M. Van der Schaar, "Delay-sensitive resource management in multi-hop cognitive radio networks," in *Proc.* IEEE DySPAN, 2008, pp. 1–12.
26. A. Qayyum, L. Viennot, and A. Laouiti "Multipoint relaying for flooding broadcast messages in mobile wireless networks," in *Proc.* HICSS, 2002, pp. 3866–3875.
27. Z. Haas, J. Halpern, and L. Li "Gossip-based ad hoc routing," *IEEE/ACM Trans. Networking*, vol. 14, no. 3, pp. 479–491, June 2006.
28. R. Gandhi, A. Mishra, and S. Parthasarathy "Minimizing broadcast latency and redundancy in ad hoc networks," *IEEE/ACM Transactions on Networking*, vol. 16, no. 4, pp. 840–851, 2008.
29. Y. Song, J. Xie, and X. Wang "A novel unified analytical model for broadcast protocols in multi-hop cognitive radio ad hoc networks," *IEEE Transactions on Mobile Computing*, 2013.
30. S. Alagar, S. Venkatesan, and J. Cleveland "Reliable broadcast in mobile wireless networks," in *Proc.* IEEE MILCOM, 1995, pp. 236–240.
31. J. J. Garcia-Luna-Aceves and Y. X. Zhang, "Reliable broadcasting in dynamic networks," in *Proc.* IEEE ICC, 1996, pp. 1630–1634.
32. M. Impett, M. S. Corson, and V. Park "A receiver-oriented approach to reliable broadcast ad hoc networks," in *Proc.* IEEE Wireless Communications and Networking Conference (WCNC), 2000, pp. 117–122.
33. Z. Chen, C. Qiao, J. Xu, and T. Lee "A constant approximation algorithm for interference aware broadcast in wireless networks," in *Proc.* IEEE INFOCOM, 2007, pp. 740–748.
34. S. Huang, P.-J. Wan, X. Jia, H. Du, and W. Shang "Broadcast scheduling in interference environment," *IEEE Transactions on Mobile Computing*, vol. 7, no. 11, pp. 1338–1348, 2008.
35. R. Mahjourian, F. Chen, R. Tiwari, M. Thai, H. Zhai, and Y. Fang "An approximation algorithm for conflict-aware broadcast scheduling in wireless ad hoc networks," in *Proc.* ACM MobiHoc, 2008, pp. 331–340.
36. R. Ramaswami and K. Parhi, "Distributed scheduling of broadcasts in a radio network," in *Proc.* IEEE INFOCOM, 1989, pp. 497–504.
37. N. Alon, A. Bar-Noy, N. Linial, and D. Peleg "A lower bound for radio broadcast," *Journal of Computer System Science*, vol. 43, pp. 290–298, October 1991.
38. B. Chlebus, L. Gasieniec, A. Gibbons, A. Pelc, and W. Rytter "Deterministic broadcasting in unknown radio networks," in *Proc.* ACM-SIAM Symposium on Discrete Algorithms (SODA), 2000, pp. 861–870.
39. B. Williams and T. Camp, "Comparison of broadcasting techniques for mobile ad hoc networks," in *Proc.* ACM International Symposium on Mobile Ad Hoc Networking & Computing (MobiHoc), 2002, pp. 194–205.
40. A. Czumaj and W. Rytter, "Broadcasting algorithms in radio networks with unknown topology," *Journal of Algorithms*, vol. 60, pp. 115–143, August 2006.
41. W. Lou and J. Wu, "Toward broadcast reliability in mobile ad hoc networks with double coverage," *IEEE Transactions on Mobile Computing*, vol. 6, no. 2, pp. 148–163, 2007.
42. C. Campolo, A. Molinaro, A. Vinel, and Y. Zhang "Modeling prioritized broadcasting in multichannel vehicular networks," *IEEE Transactions on Vehicular Technology*, vol. 61, pp. 687–701, February 2012.

43. X. Ma, J. Zhang, X. Yin, and K. S. Trivedi "Design and analysis of a robust broadcast scheme for VANET safety-related services," *IEEE Transactions on Vehicular Technology*, vol. 61, pp. 46–61, January 2012.
44. Q. Yang, J. Zheng, and L. Shen "Modeling and performance analysis of periodic broadcast in vehicular ad hoc networks," in *Proc*. IEEE GLOBECOM, 2011, pp. 1–5.
45. J. Chen "AMNP: ad hoc multichannel negotiation protocol with broadcast solutions for multi-hop mobile wireless networks," *IET Communications*, vol. 4, no. 5, pp. 521–531, 26 2010.
46. Y. Wan, X. Chen, and J. Lu "Broadcast enhanced cooperative asynchronous multichannel MAC for wireless ad hoc network," in *Proc*. WiCOM, 2011, pp. 1–5.
47. L. Lin, W. Jia, and W. Lu "Performance analysis of IEEE 802.16 multicast and broadcast polling based bandwidth request," in *Proc*. IEEE WCNC, 2007, pp. 1854–1859.
48. J. Qadir, C. T. Chou, A. Misra, and J. G. Lim "Minimum latency broadcasting in multiradio, multichannel, multirate wireless meshes," *IEEE Transactions on Mobile Computing*, vol. 8, no. 11, pp. 1510–1523, November 2009.

Chapter 2
QoS-based Broadcast Protocol Under Blind Information in Cognitive Radio Ad Hoc Networks

2.1 Network Model

In this book, we consider a CR ad hoc network where N SUs and K PUs co-exist in an $L \times L$ area, as shown in Fig. 2.1. PUs are distributed within the area under the probability density function (pdf) $f_G(g)$. For simplicity, in this book, we consider that PUs are evenly distributed. The SUs opportunistically access M licensed channels. In Fig. 2.1, the solid circle represents the transmission range of a SU with a radius of r_c. Other SUs within the transmission range are considered as the neighboring nodes of the corresponding SU. That is, only when a SU receiver is within the transmission range of a SU transmitter, the signal-to-noise ratio (SNR) at the SU receiver is considered to be acceptable for reliable communications. In addition, the dashed circle represents the sensing range of a SU with a radius of r_s. That is, if a PU is currently active within a sensing range, the corresponding SU is able to detect its presence. Since the sensing ranges of different SUs at different locations may include different PUs, their acquired available channels may be different [1–3]. In addition, because the available channels of a SU are obtained based on the sensing outcome within the sensing range, each SU is not allowed to communicate with other SUs outside its sensing range since it may mistakenly use an occupied channel by a PU, which results in interference to the PU. Therefore, in this book, we assume that $r_c \leq r_s$. Additionally, we assume that a time-slotted system is adopted for SUs [4], where the length of a slot is long enough to transmit a broadcast packet.

In addition, in this book, we model the PU channel activity as an ON/OFF process, where the length of the ON period is the length of a PU packet. We assume that each PU randomly selects a channel from the spectrum band to transmit a packet. Therefore, the packets on the same channel do not necessarily belong to the same PU. This is a more practical scenario, as compared to some papers which assume that each channel is associated with a different PU. Under such practical scenario, the number of active PUs is not necessarily the number of occupied channels but depends on the total number of PUs in the network and the PU traffic intensity.

© The Author(s) 2014
Y. Song, J. Xie, *Broadcast Design in Cognitive Radio Ad Hoc Networks,*
SpringerBriefs in Electrical and Computer Engineering, DOI 10.1007/978-3-319-12622-7_2

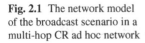

Fig. 2.1 The network model
of the broadcast scenario in a
multi-hop CR ad hoc network

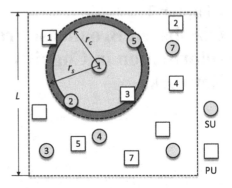

2.2 Exploring Broadcast Design in CRNs

To explore the broadcast design in CR ad hoc networks, we first investigate two
straightforward broadcast schemes in multi-hop CR ad hoc networks under blind
information. We observe that both broadcast schemes have drawbacks which make
them unsuitable to be used in CR ad hoc networks. In the rest of the book, we use the
term "sender" to indicate a SU source node or a SU who has just received a message
and will rebroadcast the message. In addition, we use the term "receiver" to indicate
a SU who has not received the message.

2.2.1 Random Broadcast Scheme

The first broadcast scheme is called the *random broadcast scheme*. Since a SU is
unaware of the channel availability information of other SUs before broadcasts are
executed, a straightforward action for a SU sender is to randomly select a channel
from its available channel set and broadcasts a message on that channel in a time slot.
In addition, as stated in Chap. 1, each SU sender needs to broadcast the message
for multiple time slots. We denote the number of time slots that each SU sender
broadcasts as S. Accordingly, for a SU receiver, without the channel availability
information of the sender, it cannot constantly stay on one channel during the whole
broadcast procedure since this channel may not be in the available channel set of the
sender, which leads to a definite failure of the broadcast. Thus, the only fair action
for the receiver is to randomly select an available channel to listen in each time slot.
If the channel selected by the receiver is the same as the channel selected by the
sender, the broadcast message can be successfully received. This broadcast scheme
is easy to be implemented in CR ad hoc networks under blind information. However,
it cannot guarantee channel rendezvous (i.e., the sender and the receiver stay on the
same channel at the same time and establish a link). In other words, in each time
slot, the sender tries its luck to broadcast to its neighboring nodes. Clearly, when
the number of channels is large, since the probability that the sender and receiver

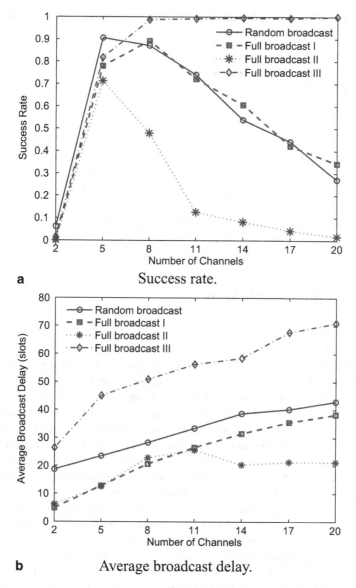

a Success rate.

b Average broadcast delay.

Fig. 2.2 Success rate and average broadcast delay of the random and full broadcast schemes under different number of channels when $N = 9$, $K = 20$, and $S = 20$

select the same channel is low, the probability that a broadcast is successful under the random broadcast scheme is fairly low.

Figure 2.2 shows the simulation results of the random broadcast scheme under different number of channels when $N = 9$, $K = 20$, and $S = 20$. The SUs form a 3×3 grid network. We assume that the PU traffic is discrete-time, where the PU packet inter-arrival time X follows the biased-geometric distribution [5]. Additionally, other parameters are listed as follows:

Side length of the simulation area $L = 10$ (unit length);
Radius of the sensing range $r_s = 2$ (unit length);
Radius of the transmission rage $r_c = 2$ (unit length);
The normalized PU arrival rate $\lambda_p = 0.5$;
The PU packet length $L_p = 10$ (time slots).

As mentioned in Chap. 1, the success rate is defined as the probability that all nodes in a network successfully receive the broadcast message and the average broadcast delay is defined as the average duration from the moment a source node starts a broadcast until the moment the last node in the network receives the broadcast message. It is shown in Fig. 2.2 that the random broadcast scheme leads to very low success rate when the number of channels is large, which is not suitable to be used in multi-hop CR ad hoc networks when the number of channels is large.

2.2.2   Full Broadcast Scheme

The second broadcast scheme is called the *full broadcast scheme* under which each SU visits all the available channels in the spectrum. Unlike the random broadcast scheme where the channel in each time slot is randomly selected by a SU, in the full broadcast scheme, a SU sender broadcasts on all its available channels sequentially. Similarly, a SU receiver listens to its available channels sequentially. In addition, we use three different channel hopping sequences for the full broadcast scheme: (1) the channel hopping sequence under which the order for each SU to visit all the available channels is random (denoted as Full broadcast I); (2) the channel hopping sequence under which each SU visits all the available channels sequentially (denoted as Full broadcast II); and (3) the jump-stay channel hopping sequence [6] (denoted as Full broadcast III). The jump-stay channel hopping sequence can be constructed under blind information with guaranteed rendezvous. Furthermore, similar to the random broadcast scheme, each SU sender also broadcasts for a finite number of time slots, S.

Figure 2.2 shows the simulation results of the full broadcast scheme using different channel hopping sequences under different number of channels when $N = 9$, $K = 20$, and $S = 20$. Compared with the random broadcast scheme, the full broadcast scheme using the first two channel hopping sequences also suffers a low success rate when the number of channels is large. This is because that these channel hopping sequences in the full broadcast scheme also cannot guarantee channel rendezvous. Moreover, the Full broadcast II scheme leads to an extremely low success rate when the number of channels is large, as compared to the Full broadcast I scheme. In addition, both the random broadcast scheme and the full broadcast scheme using the first two hopping sequences have a long average broadcast delay when the number of channels is large. On the other hand, the Full broadcast III scheme leads to a high success rate, as compared to other schemes. However, from Fig. 2.2b, this scheme has an extensively long average broadcast delay (almost as twice as the average broadcast delay in other scenarios). Hence, it is not suitable for broadcast scenarios where short broadcast delay is often required. Due to the low success rate in Full broadcast II and the long broadcast delay in Full broadcast III, in the rest of the book, we only use the random broadcast scheme and the Full broadcast I as the benchmarks to compare with our proposed schemes.

From the above discussion, it is known that these straightforward broadcast schemes have limitations to be used in multi-hop CR ad hoc networks. By investigating these broadcast schemes, we gain two useful insights for designing an efficient broadcast protocol in multi-hop CR ad hoc networks. First of all, from Fig. 2.2, it is shown that the first three schemes suffer a very low success rate when the number of channels is large because these schemes cannot guarantee channel rendezvous. Thus, a channel hopping sequence that can guarantee channel rendezvous without the channel availability information of other SUs is required to achieve a high success rate. Secondly, all these broadcast schemes are quite costly in terms of the average broadcast delay when the number of channels is large, which is not desirable for efficient broadcasts. This is because that a SU needs to use all the available channels in the spectrum for broadcasting in these schemes. If a SU only uses a subset of its available channels for broadcasting, the broadcast delay may be reduced. However, since fewer channels are used, the success rate may also be affected. Therefore, given that the success rate is not sacrificed, properly reducing the number of channels for broadcasting can result in shorter average broadcast delay.

2.3   The Basic Scheme

In this chapter, we present the basic scheme of our proposed protocol. As mentioned in Chap. 2.2, the straightforward broadcast schemes are not suitable for CR ad hoc networks. Therefore, based on the insights that we gain from these schemes, the main idea of our proposed QB^{2}IC protocol is to intelligently design the channel hopping sequences for both the SU sender and the SU receiver to guarantee channel rendezvous, given that the sender and the receiver have at least one channel in common. In addition, the SU sender broadcasts on a subset of its available channels in order to reduce the average broadcast delay.

2.3.1   The Single-Hop Scenario

First of all, we consider the single-hop broadcast scenario. We propose a novel channel hopping strategy for SUs to guarantee channel rendezvous. There are several existing work on single-hop channel rendezvous for CR networks [6–9]. A common feature of these prior proposals is that all SUs in a network have to follow the same mechanism to construct the channel hopping sequence for rendezvous regardless

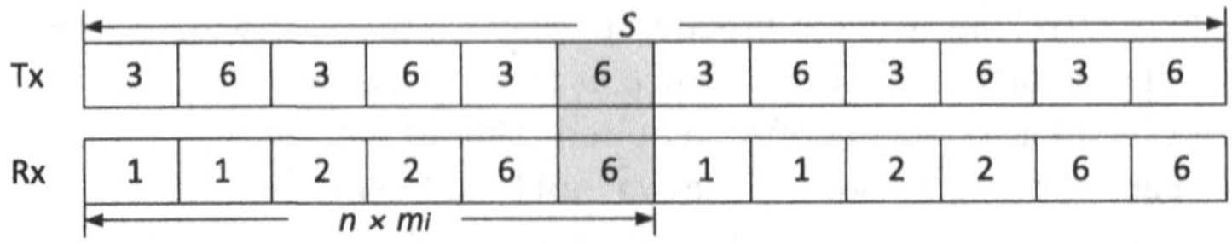

Tx	3	6	3	6	3	6	3	6	3	6	3	6
Rx	1	1	2	2	6	6	1	1	2	2	6	6

$\longleftarrow$ S $\longrightarrow$

$\longleftarrow$ $n \times mi$ $\longrightarrow$

Fig. 2.3 An example of the QoS-based broadcast protocol

of transmitters or receivers. However, as stated in Chap. 1, in [7–9], the proposed channel hopping schemes cannot guarantee channel rendezvous in all scenarios under blind information. In [6], even though the proposed channel hopping sequence in the asymmetric model can guarantee channel rendezvous within $6MP(P - G)$ time slots, where P is the smallest prime number larger than M and G is the number of common channels between two SUs, $6MP(P - G)$ is usually a very large number when M is large. Therefore, this scheme may lead to very long broadcast delay when M is large. In fact, the communication pair can follow different mechanisms to construct the channel hopping sequences for channel rendezvous, which is ignored in all prior proposals. Thus, in this book, we use the channel hopping sequences generated by different methods for the sender and the receiver to guarantee channel rendezvous. Compared with the previous proposals in [6–9], our proposed channel hopping sequences can guarantee rendezvous in all scenarios under blind information within M^2 time slots, which is more favorable in broadcast scenarios. Under our proposed basic QB^{2}IC scheme, a SU sender first randomly selects n channels from its available channel set. Then, it hops and broadcasts periodically on the selected n channels for S time slots. This channel hopping sequence with a length of S time slots is named as *the broadcast sequence*. The values of n and S are determined by the QoS requirements of the network. On the other hand, for each receiver, it first forms a random sequence that consists of its every available channel with a length of n time slots for each channel, namely *the receiving sequence*. Then, it hops and listens following the receiving sequence periodically. Denote the number of available channels of SU_i as m_i. Hence, the length of the receiving sequence of SU_i is $n \times m_i$. Figure 2.3 shows an example of the proposed QoS-based broadcast protocol. If the available channel set of the sender is $\{1, 2, 3, 6\}$, it randomly selects two channels (e.g., $\{3, 6\}$) to broadcast for S slots (i.e., $n = 2$). Each receiver listens for two time slots on each available channel of its available channel set (e.g., $\{1, 2, 6\}$) periodically. Thus, if $S = 12$, the broadcast sequence for the sender is $\{3, 6, 3, 6, 3, 6, 3, 6, 3, 6, 3, 6\}$. In addition, the receiving sequence for the receiver is $\{1, 1, 2, 2, 6, 6, \cdots\}$. Hence, a successful broadcast is performed when both SUs hop on channel 6.

Based on the above rules, if the SU sender selects all its available channels and the length of the broadcast sequence S is equal to $n \times M$, the channel rendezvous is guaranteed within S time slots when the sender and each receiver have at least one channel in common. Therefore, the broadcast is ensured to be successful in the single-hop scenario. Thus, if n is sufficiently large, the probability that at least one channel selected by the sender are also in the available channel set of each receiver is high. However, on the other hand, since each SU is only equipped with one radio, it

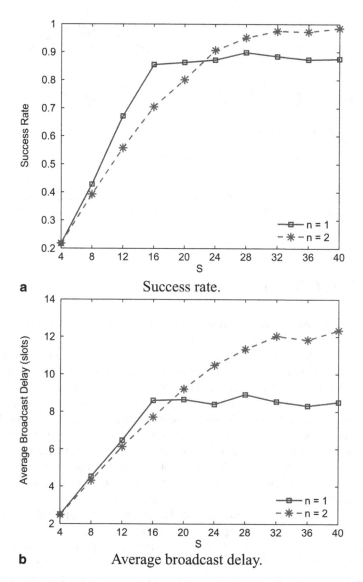

a   Success rate.

b   Average broadcast delay.

Fig. 2.4 The trade-off between the success rate and average broadcast delay under different n and S

cannot broadcast on n $(n > 1)$ channels simultaneously. Hence, it takes a long time to finish broadcasting on all n channels when n is large.

From the above analysis, there exists a trade-off between the success rate and the average broadcast delay for different values of n and S. Figure 1 shows the simulation results of the success rate and the average broadcast delay for a single-hop broadcast scenario when $K = 40$ under various values of n and S. The PU traffic and other parameters are the same to generate Fig. 2.2. It is illustrated that higher success rate

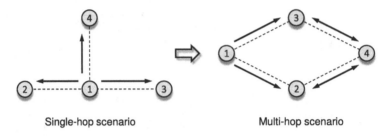

Single-hop scenario   Multi-hop scenario

Fig. 2.5 Two different topologies of two 4-SU networks

often indicates longer average broadcast delay. Therefore, if the QoS requirements (i.e., the minimum required success rate and the maximum allowed average broadcast delay) are given, proper n and S can be selected.

2.3.2   The Multi-Hop Scenario

Next, we investigate the basic scheme of the proposed protocol in the multi-hop broadcast scenario. We consider two 4-SU networks as shown in Fig. 2.5. The first topology of the 4-SU network is a single-hop scenario where SU_1 is the source node. The second topology is a multi-hop scenario evolved from the first topology when SU_1 moves away from SU_4. Since we want to consider the broadcast collision issue, the multi-hop scenario is studied in such a grid topology instead of a simple chain topology, as shown in Fig. 2.5. Without loss of generality, we assume that SU_1 is the source node and other nodes are by default receivers. Each receiver first follows the proposed receiving sequence to hop through and listen on the channels. When a receiver successfully receives the broadcast message, it becomes a sender who needs to rebroadcast the message. Therefore, it follows the rules of the senders and generates the proposed broadcast sequence to rebroadcast.

Our proposed QB^2IC protocol has special advantages when applied to multi-hop scenarios in CR networks, as compared to traditional ad hoc networks. In traditional ad hoc networks, if a node (e.g., SU_4) receives multiple copies of a message from its parent nodes (e.g., SU_2 and SU_3) simultaneously, a broadcast collision occurs and all copies of the message are discarded. Unfortunately, in traditional ad hoc networks, such broadcast collision is unavoidable in multi-hop scenarios if the parent nodes broadcast at the same time. However, under our proposed QB^2IC protocol, since the receiver can only listen to one channel at a time, as long as the parent nodes do not select the same channel to broadcast, such broadcast collision can be avoided. In fact, when the number of channels is large and different SUs obtain different available channels, the probability that two parent nodes select the same channel at the same time is fairly low.

In addition, under our proposed QB^2IC protocol, the success rate of the broadcast for the whole network can be improved in the multi-hop scenario. This is because that a SU may not only receive the broadcast message from its parent node, but also receive the message from its child node (e.g., SU_2 can receive the message from SU_4 if SU_4 receives the message from the path $SU_1 \rightarrow SU_3 \rightarrow SU_4$). This is usually different from the broadcast schemes in traditional MANETs where nodes receive broadcast messages from their parent nodes. More importantly, if the channels used for broadcast in different paths are different, the probability that one of the channels is in the available channel set of the receiver is increased, as compared to the scenario where a SU only receives the message from its parent nodes. Thus, by utilizing the diversity of users and channels, the success rate of the whole network can be increased. Figure 2.6 shows the performance comparison between the single-hop scenario and the multi-hop scenario shown in Fig. 2.5 when $n = 1$. It is illustrated that the improvement of the success rate under the multi-hop scenario is up to 30 %, while the multi-hop scenario only costs up to 20 % additional average broadcast delay. Therefore, under our proposed QB^{2}IC protocol, the success rate is benefited in the multi-hop scenario due to the diversity of users and channels.

2.4 The Enhanced Scheme

In this chapter, we first conduct an analysis on the channel availability of different SUs. Then, based on the results of this analysis, an enhanced QB^{2}IC scheme for multi-hop CR ad hoc networks is presented.

2.4.1 Analysis of the Channel Availability

Based on the considered network model, the available channels of a SU are determined by the active PUs within its sensing range. Thus, first of all, we derive the average number of available channels of a SU. The size of the simulation area and the sensing range is denoted as A_L and A_S, respectively. Since PUs are evenly distributed in the considered simulation area, the probability that p PUs are in a sensing range is

$$\Pr(p) = \binom{K}{p} \left(\frac{A_S}{A_L} \right)^p \left(\frac{A_L - A_S}{A_L} \right)^{K-p}, \tag{2.1}$$

where $\binom{K}{p}$ represents the total combinations of K choosing p. In addition, we denote the probability that a PU is active as ρ. Therefore, given that there are p PUs in a sensing range, the probability that there are b PUs active is

$$\Pr(b|p) = \binom{p}{b} \rho^b (1 - \rho)^{p-b}. \tag{2.2}$$

Furthermore, given that there are p PUs and b active PUs within a sensing range, the probability that there are c available channels is denoted as $\Pr(c|p, b)$. Since

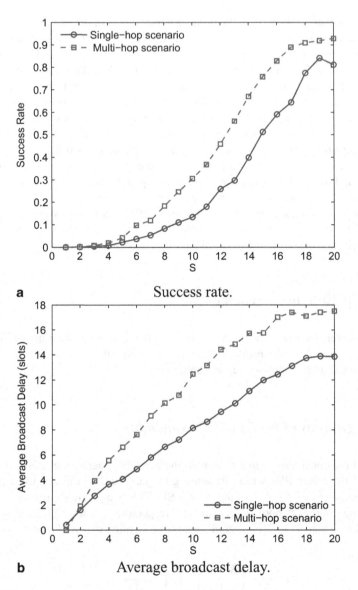

Fig. 2.6 Comparison between the single-hop and the multi-hop scenarios

the number of available channels is only related to the number of active PUs, c is independent of p. In addition, since an active PU randomly selects a channel from M channels in the spectrum, $\Pr(c|p, b)$ is equivalent to the probability that there are exactly c empty boxes given that b distinguishable balls are randomly put into a total of M distinguishable boxes and a box can have more than one ball (because we do

Fig. 2.7 Two neighboring
SUs whose sensing ranges
overlap

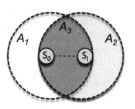

not limit a channel to only one PU). Thus, $\Pr(c|p,b)$ can be expressed as:

$$\Pr(c|p,b) = \frac{\binom{M}{c}(M-c)!\,S(b,M-c)}{M^b}, c \in [\max(0, M-b), M], \qquad (2.3)$$

where $S(b, M-c)$ is the Stirling number of the second kind. In addition, $S(b, M-c)$ is defined as

$$S(b, M-c) = \frac{1}{(M-c)!}\sum_{i=0}^{M-c}(-1)^i\binom{M-c}{i}(M-c-i)^b. \qquad (2.4)$$

Thus, the probability that there are c available channels and there are p PUs and b active PUs in the sensing range of a SU is the product of (2.1), (2.2), and (2.3). Then, the average number of available channels of a SU, $E[c]$, is written as

$$E[c] = \sum_{p=0}^{K}\sum_{b=0}^{p}\sum_{c=\max(0,M-b)}^{M} \frac{c\binom{M}{c}(M-c)!\,S(b, M-c)}{M^b}$$
$$\binom{p}{b}\rho^b(1-\rho)^{p-b}\binom{K}{p}\left(\frac{A_S}{A_L}\right)^p\left(\frac{A_L-A_S}{A_L}\right)^{K-p}. \qquad (2.5)$$

In addition, another important parameter is the average number of common channels between two neighboring SUs. Figure 2.7 illustrates an example of two neighboring SUs whose sensing ranges overlap, where d is the distance between the two SUs. Assume that SU_i and SU_j can hear each other. As shown in Fig. 2.7, the sensing ranges of the two SUs are divided into three areas. A_3 (the dark area) represents the area of the overlapping part, while A_1 (the white area) and A_2 (the gray area) represent the areas of the sensing ranges of SU_i and SU_j without A_3, respectively.

Define $A^* = A_1 + A_2 + A_3$. Therefore, based on basic geometry, A^* can be obtained as follows:

$$A^* = (2\pi - 2\alpha)r_s^2 + d\sqrt{r_s^2 - \left(\frac{d}{2}\right)^2}, \qquad (2.6)$$

where $\alpha = \cos^{-1}\frac{d}{2r_s}$. Thus, the channels that are used by the active PUs within A^* are those that cannot be used by either SU_i or SU_j. In other words, the common channels between two neighboring SUs are those that are not used by the active PUs

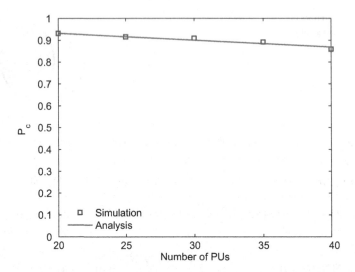

Fig. 2.8 The ratio of the average number of common channels between two neighboring SUs to the average number of available channels of one SU when $\rho = 0.9$

within A^*. Thus, similar to the derivation process for $E[c]$, the average number of common channels between two neighboring SUs, $E[u]$, is obtained from

$$E[u] = \sum_{y=0}^{K} \sum_{z=0}^{y} \sum_{u=\max(0,M-z)}^{M} \frac{u\binom{M}{u}(M-u)!S(z,M-u)}{M^z}$$
$$\binom{y}{z}\rho^z(1-\rho)^{y-z}\binom{K}{y}\left(\frac{A^*}{A_L}\right)^y\left(\frac{A_L - A^*}{A_L}\right)^{K-y}, \qquad (2.7)$$

where y and z are the number of PUs and active PUs within A^*, respectively.

Then, we define the ratio of the average number of common channels between two neighboring SUs to the average number of available channels of a SU as

$$P_c = \frac{E[u]}{E[c]}, \qquad (2.8)$$

P_c measures the similarity of the available channel sets between two neighboring SUs. If $P_c = 1$, this means that the available channels between two neighboring SUs are exactly the same. On the other hand, if $P_c = 0$, this means that the available channels between two neighboring SUs are completely different. Figure 2.8 shows the simulation and analytical results of P_c when $r_c = r_s$ and both SUs are at the border of each other's sensing range. Since the sensed available channels of two SUs might be the most distinct when they are apart the most, Fig. 2.8 shows the lower bound of P_c. Figure 2.8 indicates that the simulation and analytical results coincide and also implies that even though the available channels of different SUs are different, the similarity of available channels between neighboring SUs is high ($> 85\%$).

2.4.2 The Enhanced QB^2IC Scheme

The above analysis of the channel availability indicates that neighboring SUs have very similar available channel sets. Thus, inspired by this observation, we propose an enhanced QB²IC scheme to further improve the performance.

The main idea of our enhanced scheme is that each SU selects the first θ channels from its available channel set based on the indexes of the channels to form a new available channel set. Based on the downsized available channel set, each SU follows the basic scheme to broadcast. Since the available channel sets of neighboring SUs are similar, the downsized available channel sets of neighboring SUs are also similar. In this way, if the threshold θ is properly selected, the success rate does not degrade significantly. However, since the number of the channels that each receiver needs to listen is reduced, the average broadcast delay can be greatly reduced.

However, there again exists a trade-off between the success rate and the average broadcast delay when selecting the threshold θ. As shown in Fig. 2.9, when θ increases, the success rate increases but the average broadcast delay also increases. Thus, if the QoS requirements are given, a proper θ can be selected.

2.5 Performance Evaluation

In this chapter, we evaluate the performance of the proposed QB²IC protocol. Since there is no existing comparable broadcast scheme under blind information for multi-hop CR ad hoc networks, we compare our proposed broadcast schemes with the random broadcast scheme and the full broadcast scheme with the first channel hopping sequence introduced in Chap. 2.2. As mentioned in Chap. 2.2, we also assume that the PU traffic is discrete-time, where the PU packet inter-arrival time X follows the biased-geometric distribution whose probability mass function (pmf) is given by [10, 11]:

$$\Pr(X = x) = \begin{cases} 0 & x < l \\ \lambda_p (1 - \lambda_p)^{(x-l)} & x \geq l, \end{cases} \tag{2.9}$$

where x is the number of time slots between packet arrivals, $l \geq 0$ represents the minimum number of time slots between two adjacent packets, and λ_p is the probability that a PU packet arrives during one time slot (i.e., λ_p is the normalized arrival rate of PU packets). Thus, the probability that a PU is active can be written as $\rho = \frac{L_p}{L_p + \frac{1-\lambda_p}{\lambda_p}}$, where L_p is the fixed PU packet length [12]. It is noted that the PU traffic model is used to obtain simulation results. In fact, our proposed QB²IC broadcast protocol does not rely on specific PU traffic models. In addition, denote σ as the probability that a transmission is successful. That is, if $\sigma = 1$, it means that there is no transmission error and a transmission is always successful. Moreover, the default parameters used to obtain the simulation results are listed in Table 2.1. For the topology of the

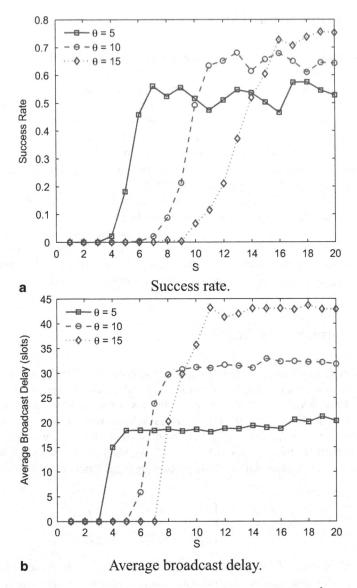

Fig. 2.9 Success rate and average broadcast delay of the proposed enhanced QB²IC scheme under various θ

CR ad hoc network, we assume that SUs form a 4×4 grid network with the distance between two adjacent SUs equal to r_c.

Figure 2.10 depicts the performance results of the two proposed QB²IC schemes (for the enhanced scheme, $\theta = 10$), the random broadcast scheme, and the full broadcast scheme under different S when $K = 40$, $M = 20$, $n = 1$, and $\sigma = 1$. It is shown that when S is large (e.g., $S = 19$), the basic scheme outperforms the other

Table 2.1 Simulation parameters

Number of SUs N	16
Number of PUs K	40
Number of channels M	20
Side length of the simulation area L	10 (unit length)
Radius of the sensing range r_s	2 (unit length)
Radius of the transmission rage r_c	2 (unit length)
Number of selected channels n	1
The normalized PU arrival rate λ_p	0.5
The PU packet length L_p	10 (time slots)
The probability of a successful transmission σ	1

three broadcast schemes in terms of higher success rate. However, the enhanced scheme can greatly reduce the average broadcast delay while obtaining satisfactory success rate. Both the proposed QB^{2}IC schemes outperform the random broadcast scheme and the full broadcast scheme in terms of higher success rate and shorter average broadcast delay.

Figures 2.11 and 2.12 show the impact of the number SUs and the number of PUs on the network performance, respectively, where the secondary network is a grid network when $S = 20$. On the other hand, Fig. 2.13 depicts the impact of the number of SUs on the network performance where the secondary network is not a grid network. In Fig. 2.11, SUs form a 2×2 network for $N = 4$ and a 3×3 network for $N = 9$. It is shown that the average broadcast delay increases as the number of SUs in the network increases. This is because that the increase of the number of hops leads to a longer broadcast delay. However, the success rate of the basic QB^{2}IC scheme does not decrease significantly when the number of SUs increases. This is because that the multi-hop scenario benefits the success rate due to the diversity of channels. On the other hand, as shown in Fig. 2.12, the success rate decreases when the number of PUs increases. This is because that the number of common channels between neighboring nodes decreases when the total number of PUs increases, which leads to a lower success rate.

We also investigate the impact of the number of SUs on the network performance where the secondary network is not a grid network. Fig. 2.13 shows the results of the basic broadcast scheme where the SUs are randomly distributed in the simulation area. In Fig. 2.13a, it is shown that when the number of SUs increases, the success rate of the whole network also increases. This is because that if the number of SUs within the same area is large, the diversity of senders and channels increases. Thus, the number of potential senders of a SU increases. Therefore, in our proposed QB^{2}IC protocol, the probability that one of the channels used by these senders is in the available channel set of the receiver is increased. Hence, the probability that all nodes in the network can successfully receive the broadcast message also increases. Due to the same reason, in Fig. 2.13b, it is shown that the average broadcast delay does not increase significantly when the number of SUs increases. That is, when the

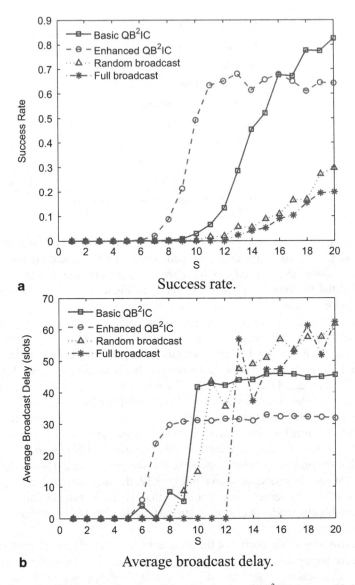

a   Success rate.

b   Average broadcast delay.

Fig. 2.10 Success rate and average broadcast delay of the proposed QB^2IC schemes with the random and full broadcast schemes under various S

number of SUs increases by 10 times, the average broadcast delay only increases by 40 % for the basic scheme and 51 % for the enhanced scheme. To sum up, it is shown that our proposed basic and enhanced broadcast schemes are scalable when the CR ad hoc network size increases.

Figure 2.14 shows the impact of the number of channels on the network performance when $K = 30$ and $N = 9$. For the length of the broadcasting sequence, we

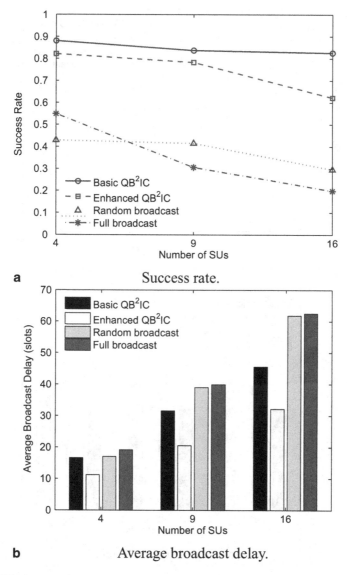

a Success rate.

b Average broadcast delay.

Fig. 2.11 The impact of the number of SUs on the network performance where the secondary network is a grid network

set $S = 2M$. In addition, for the enhanced scheme, we let $\theta = M/2$. As stated in Chap. 2.3, the number of channels leads to a trade-off in terms of the success rate and broadcast delay. On one hand, a large M ensures that the probability that two parent nodes select the same channel at the same time is low. Therefore, as shown in Fig. 33, expect the full broadcast scheme, the success rate of the other three broadcast schemes increases as the number of channels increases. In addition, both

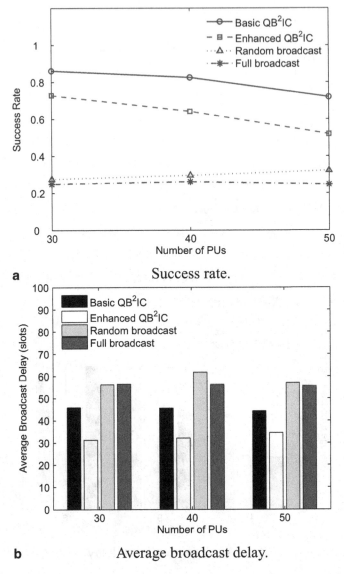

a   Success rate.

b   Average broadcast delay.

Fig. 2.12 The impact of the number of PUs on the network performance

our proposed basic and enhanced schemes have very similar and high success rates (i.e., > 0.95). However, the random and full broadcast schemes result in relatively low success rates. On the other hand, a large M also leads to a long broadcast delay. Hence, as shown in Fig. 2.14b, the average broadcast delay of all four schemes increases as the number of channels increases. Moreover, our proposed enhanced scheme outperforms other the three broadcast schemes in terms of shorter average broadcast delay.

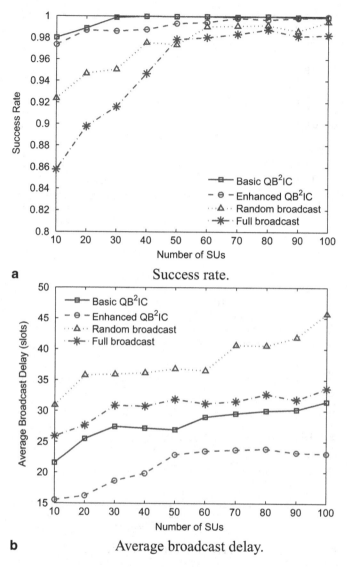

Fig. 2.13 The impact of the number of SUs on the network performance where the secondary network is *not* a grid network

In this chapter, we study the impact of the channel availability variation on the network performance. Note that during a broadcast process, the channel availability of the sender and receiver may change due to either PU activity (i.e., a new PU may claim one of the available channels during a broadcast) and PU mobility (i.e., a new active PU may move into the sensing range of a SU). This channel availability change may also affect the network performance. Therefore, the robustness of the proposed

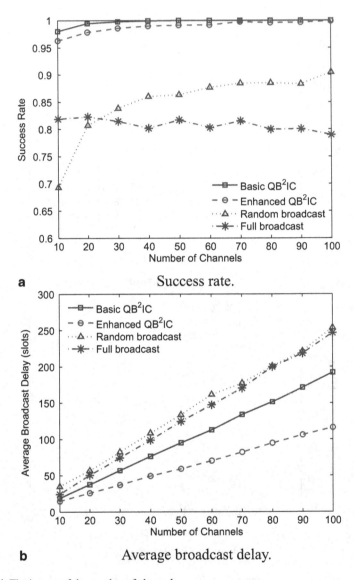

a   Success rate.

b   Average broadcast delay.

Fig. 2.14 The impact of the number of channels

broadcast scheme against the channel availability variation needs to be investigated. Figure 2.15 depicts the impact of the channel availability change due to PU activity on the basic broadcast scheme when $n = 2$, $S = 40$ and $M = 20$. We let the PU packet arrival rate change from 5 to 50 pkt/s and the PU packet length is 50 slots. This means that the probability that a PU is active, ρ, varies from 0.35 to 0.85. It is shown that the impact of the PU activity on the network performance is quite limited. When PUs are very densely deployed (i.e., 40 PUs within a 10×10 area) and PU

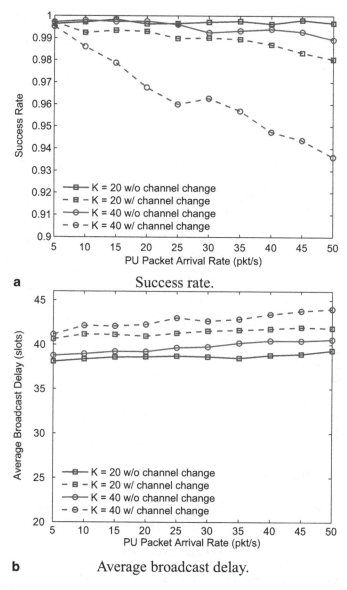

a   Success rate.

b   Average broadcast delay.

Fig. 2.15 The impact of the channel availability change due to PU activity

traffic is very heavy (i.e., $\rho = 0.85$), the decrease of the success rate is only up to 5 % and the increase of the average broadcast delay is only up to 9 %.

In addition, Fig. 2.15 depicts the impact of the channel availability change due to PU mobility. We use the Random Waypoint Mobility Model to characterize the movement pattern for PUs [13–15]. Under this model, a PU begins by staying in a location for a certain period of time (e.g., a pause time which is uniformly distributed

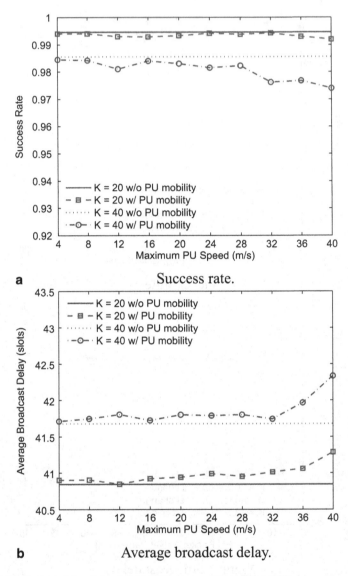

a   Success rate.

b   Average broadcast delay.

Fig. 2.16 The impact of the channel availability change due to PU mobility

between $[0, 4s]$). Once the pause period expires, the PU randomly selects a speed that is uniformly distributed between $[0, v_{max}]$ and moves for a random time that is uniformly distributed between $[0, 4s]$. Upon its arrival, the PU pauses for a random time period before starting the movement again. Figure 2.16 depicts the impact of the channel availability change due to PU mobility on the network performance. Similar to the impact of PU activity on the network performance, it is shown that the impact of the PU mobility is also quite trivial. When PUs are very densely deployed

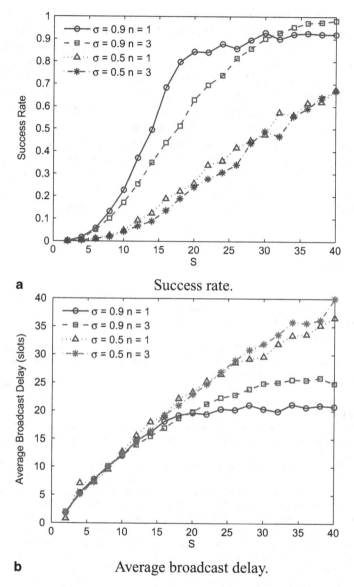

a   Success rate.

b   Average broadcast delay.

Fig. 2.17 Success rate and average broadcast delay of the basic scheme of the proposed QB^{2}IC protocol when $N = 4$ under various n, S, and σ

(i.e., 40 PUs within a 10×10 area) and PUs move very fast (i.e., the maximum PU speed is 40 m/s), the decrease of the success rate is only up to 1 % and the increase of the average broadcast delay is only up to 2 %.

In this chapter, we investigate the impact of transmission errors on the network performance. It is known that various factors (e.g., transmission contention, channel quality, etc.) can lead to transmission errors. These transmission errors may cause failed broadcasts. However, as stated in Chap. 1, due to the ACK implosion problem, ACK messages are not feasible to be used to prevent failed broadcasts in CR ad hoc networks. Even though ACKs are not applied to solve the transmission error problem, our proposed QB^{2}IC protocol can still be used in a radio environment where transmission errors exist. By increasing the number of channels selected by the sender (i.e., n) or increasing the number of times that the sender broadcasts the message (i.e., S), the probability that a sender and a receiver have a channel rendezvous increases. Therefore, our proposed QB^{2}IC protocol can still achieve satisfactory QoS requirements. Figure 2.17 shows the simulation results of the basic scheme of the proposed QB^{2}IC protocol when $N = 4$ under various n, S, and σ, where SUs form a 2×2 grid network. It is shown that by increasing n or S, our proposed QB^{2}IC protocol can still achieve satisfactory performance.

References

1. Y. Song and J. Xie, "A QoS-based broadcast protocol for multi-hop cognitive radio ad hoc networks under blind information," in *Proc.* IEEE GLOBECOM, 2011, pp. 1–5.
2. Y. Song and J. Xie, "A distributed broadcast protocol in multi-hop cognitive radio ad hoc networks without a common control channel," in *Proc.* IEEE INFOCOM, 2012.
3. C. Gao, Y. Shi, Y. T. Hou, H. D. Sherali, and H. Zhou "Multicast communications in multi-hop cognitive radio networks," *IEEE Journal on Selected Areas in Communications (JSAC)*, vol. 29, no. 4, pp. 784–793, April 2011.
4. Y. Song and J. Xie, "Performance analysis of spectrum handoff for cognitive radio ad hoc networks without common control channel under homogeneous primary traffic," in *Proc.* IEEE INFOCOM, 2011, pp. 3011–3019.
5. Y. Song and J. Xie, "ProSpect: A proactive spectrum handoff framework for cognitive radio ad hoc networks without common control channel," *IEEE Transactions on Mobile Computing*, vol. 11, no. 7, July 2012.
6. Z. Lin, H. Liu, X. Chu, and Y.-W. Leung "Jump-stay based channel hopping algorithm with guaranteed rendezvous for cognitive radio networks," in *Proc.* IEEE INFOCOM, 2011.
7. K. Bian, J.-M. Park, and R. Chen "Control channel establishment in cognitive radio networks using channel hopping," *IEEE JSAC*, vol. 29, no. 4, pp. 689–703, April 2011.
8. N. Theis, R. Thomas, and L. DaSilva "Rendezvous for cognitive radios," *IEEE Trans. Mobile Computing*, vol. 10, no. 2, pp. 216–227, 2010.
9. C. Cormio and K. R. Chowdhury, "Common control channel design for cognitive radio wireless ad hoc networks using adaptive frequency hopping," *Ad Hoc Networks*, vol. 8, no. 4, pp. 430–438, 2010.
10. F. Gebali, *Analysis of Computer and Communication Networks*. Springer, 2008.
11. Y. Song and J. Xie, "BRACER: A distributed broadcast protocol in multi-hop cognitive radio ad hoc networks with collision avoidance," *IEEE Transactions on Mobile Computing*, 2014.
12. Y. Song and J. Xie, "Common hopping based proactive spectrum handoff in cognitive radio ad hoc networks," in *Proc.* IEEE GLOBECOM, 2010, pp. 1–5.
13. T. Camp, J. Boleng, and V. Davies "A survey of mobility models for ad hoc network research," *Wireless Communications and Mobile Computing*, vol. 2, pp. 483–502, August 2002.
14. Y. Song and J. Xie, "End-to-end congestion control in multi-hop cognitive radio ad hoc networks: To timeout or not to timeout?," in *Proc.* IEEE Globecom, 2013, pp. 4390–4395.
15. Y. Song "An IEEE 802.11 DCF-based optimal data link layer spectrum sensing scheme in cognitive radio networks," in *Proc.* IEEE Globecom, 2014.

Chapter 3
Distributed Broadcast Protocol with Collision Avoidance in Cognitive Radio Ad Hoc Networks

3.1 The Distributed Broadcast Protocol with Collision Avoidance

In this chapter, we introduce the proposed broadcast protocol for multi-hop CR ad hoc networks, BRACER. There are three components of the proposed BRACER protocol: (1) the construction of the broadcasting sequences; (2) the distributed broadcast scheduling scheme; and (3) the broadcast collision avoidance scheme. We assume that a time-slotted system is adopted for SUs, where the length of a time slot is long enough to transmit a broadcast packet [1]. In addition, we assume that the locations of SUs are static. We also assume that each SU knows the locations of its all 2-hop neighbors. We claim that this is a more valid assumption than the knowledge of global network topology. We provide a detailed discussion on this issue in Chap. 3.3. In the rest of the book, we use the term "sender" to indicate a SU who has just received a message and will rebroadcast the message. In addition, we use the term "receiver" to indicate a SU who has not received the message. The notations used in our protocol design are listed in Table 3.1.

3.1.1 Construction of the Broadcasting Sequences

The broadcasting sequences are the sequences of channels by which a sender and its receivers hop for successful broadcasts. First of all, we consider the single-hop broadcast scenario. As explained in Chap. 1, due to the non-uniform channel availability in CR ad hoc networks, a SU sender may have to use multiple channels for broadcasting in order to let all its neighboring nodes receive the broadcast message. Accordingly, the neighboring nodes may also have to listen to multiple channels in order to receive the broadcast message. Hence, the first issue to design a broadcast protocol is which channels should be used for broadcasting. One possible method is to broadcast on all the available channels of the SU sender. However, this method is quite costly in terms of the broadcast delay when the number of available channels is large. Therefore, we propose to select a subset of available channels from the original available channel set of each SU. First, the available channels of each SU

© The Author(s) 2014
Y. Song, J. Xie, *Broadcast Design in Cognitive Radio Ad Hoc Networks*,
SpringerBriefs in Electrical and Computer Engineering, DOI 10.1007/978-3-319-12622-7_3

Table 3.1 Notations used in the protocol

$N(v)$	The set of the neighboring nodes of node v
$N(N(v))$	The set of the neighbors of the neighboring nodes of node v
$d(v,u)$	The Euclidean distance between node v and u
r_c	The radius of the transmission range of each node
$|\cdot|$	The number of elements in a set
L_v	The downsized available channel set of node v
$w(v)$	The size of the downsized available channel set of node v
C	The set of the initial w of intermediate nodes
BS_v	The broadcasting sequence for a sender v
RS_v	The broadcasting sequence for a receiver v
DS_v	The default sequence of a sender v
st_v	The starting time slot of a sender v
rt_v	The time slot that a receiver v receives the message
R_v	The random number assigned to a receiver v by its sender

are ranked based on the channel indexes. Then, each SU selects the first w channels from the ranked channel list and forms a downsized available channel set. The value of w needs to be carefully designed to ensure that at least one common channel exists between the downsized available channel sets of the SU sender and each of its neighboring nodes. The detailed derivation process to obtain a proper w is given in Chap. 3.2. Based on the derivation process, each SU can calculate the value of w of its own and its 1-hop neighbors before a broadcast starts.

On the other hand, the second issue is the sequences of the channels by which a sender and its receivers hop for successful broadcasts. In this book, we design different broadcasting sequences for a SU sender and its receivers to guarantee a successful broadcast in the single-hop scenario as long as they have at least one common channel. The sender hops and broadcasts a message on each channel in a time slot following its own sequence. On the other hand, the receiver hops and listens on each channel following its own sequence. The pseudo-codes for constructing the broadcasting sequences are shown in Algorithm 1 and 2. $w(v)$ is the initial w of node v.

Algorithm 1: Construction of the broadcasting sequence BS_v for a SU sender v.

Input: $w(v), L_v$.
Output: BS_v.
    randomize the order of elements in L_v;
    $BS_v \leftarrow \varnothing$;   `/* initialization */`
    $i \leftarrow 1$;
    while $i \le w(v)^2$ **do**
        $BS_v(i) \leftarrow L_v((i \mod w(v))+1)$;
        $i \leftarrow i+1$;   `/* repeat` L_v `for` $w(v)$ `times */`
    return BS_v;

Algorithm 2: Construction of the broadcasting sequence RS_v for a SU receiver v.

Input: $w(v), L_v$.
Output: RS_v.
   randomize the order of elements in L_v;
   $RS_v \leftarrow \varnothing$;   `/* initialization */`
   $i \leftarrow 1$;
   while $i \leq w(v)$ **do**
      $j \leftarrow 1$;
      while $j \leq w(v)$ **do**
         $RS_v((i-1)w(v)+j) \leftarrow L_v(i)$;
         $j \leftarrow j+1$;   `/* repeat an element for w(v) times */`
      $i \leftarrow i+1$;   `/* repeat for every element in Lv */`
   return RS_v;

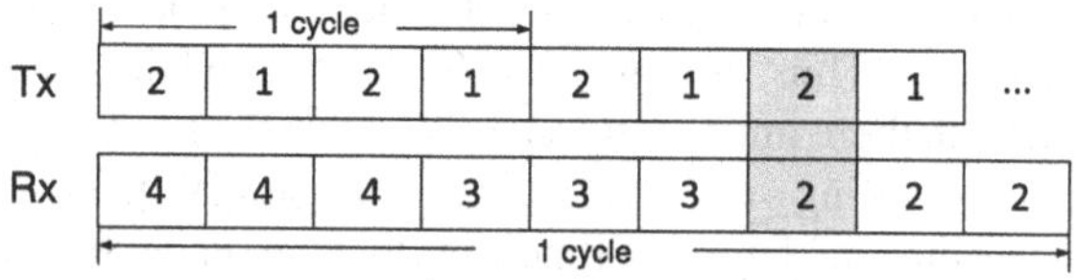

Fig. 3.1 An example of the broadcasting sequences

From Algorithm 1 and 2, for a SU sender, it hops periodically on the w available channels for w periods (i.e., w^2 time slots). For each receiver, it stays on one of the w available channels for w time slots. Then, it repeats for every channel in the w available channels. Figure 3.1 gives an example to illustrate the construction of the broadcasting sequences for SU senders and receivers. In Fig. 3.1, the downsized available channel set of a sender and a receiver is $\{1, 2\}$ and $\{2, 3, 4\}$, respectively. Based on Algorithm 2, the broadcasting sequence of the sender is $\{2, 1, 2, 1\}$. Similarly, based on Algorithm 1, the broadcasting sequence of the receiver is $\{4, 4, 4, 3, 3, 3, 2, 2, 2\}$. Since a sender usually does not know the length of the broadcasting sequence of the receiver, it broadcasts the message following its broadcasting sequence for $\lfloor \frac{M^2}{w^2} \rfloor + 1$ cycles, where M is the total number of channels. In this way, the total length of time slots that the sender broadcasts is bound to be longer than one cycle of the receiver's broadcasting sequence. As shown in Fig. 3.1, the shaded part represents a successful broadcast.

Since every SU calculates the initial value of w based on its local information and the derivation process in Chap. 3.2, different SUs may obtain different values of w. We further denote w_s and w_r as the w used by the sender and the receiver to construct their broadcasting sequences, respectively. Note that w_s and w_r may not necessarily be the same as the initial w calculated by each SU. They also depend on the initial w of its neighboring nodes. The following theorem gives an upper-bound on the single-hop broadcast delay.

Theorem 1 *If $w_s \leq w_r$, the single-hop broadcast is a guaranteed success within w_r^2 time slots as long as the sender and the receiver have at least one common channel between their downsized available channel sets.*

Fig. 3.2 A multi-hop
broadcast scenario

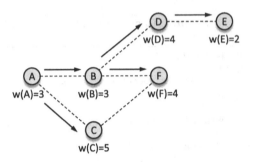

Proof Based on Algorithm 2, a SU sender broadcasts on all the channels in its downsized available channel set in w_s consecutive time slots. Based on Algorithm 1, a SU receiver listens to every channel in its downsized available channel set for w_r consecutive time slots. If $w_s \leq w_r$, during the w_r consecutive time slots for which the SU receiver stays on the same channel, every channel of the SU sender must appear at least once. Thus, as long as the SU sender and the receiver have at least one common channel, there must exists a time slot that the sender and the receiver hop on the same channel during one cycle of the broadcasting sequence of the receiver (i.e., w_r^2). Since we let the total length of time slots that the sender broadcasts be longer than one cycle of the receiver's broadcasting sequence, the broadcast is guaranteed to be successful.

Then, how to determine w_s and w_r? From Theorem 1, $w_s \leq w_r$ is a sufficient condition of a single-hop successful broadcast. Therefore, in order to satisfy this condition, a proper w_r needs to be selected by any SU who has not received the broadcast message to ensure the reception of the broadcast message sent from any potential neighbor. Since w_r depends on w_s and a SU receiver usually does not know which neighboring node is sending until it receives the broadcast message, it selects the largest initial w of all its 1-hop neighbors as its w_r. That is, for a SU receiver v, $w_r(v) = \max\{w(u)|u \in N(v)\}$. On the other hand, the sender uses its calculated initial w as w_s to broadcast. Therefore, the w_s selected by the actual sender is bound to be smaller than or equal to this w_r. Thus, according to Theorem 1, the single-hop broadcast is a guaranteed success as long as the sender and its receiver have at least one common channel between their downsized available channel sets.

To illustrate the above discussed operation, we consider a multi-hop scenario shown in Fig. 3.2. The initial w calculated by each SU before the broadcast starts based on its local information are shown. Every node also calculates the initial w of its 1-hop neighbors. Without loss of generality, node A is assumed to be the source node. Based on Theorem 1, the values of w_r employed by each receiver can be obtained. For instance, since node B knows the initial w of its neighbors (i.e., $w(A) = 3$, $w(D) = 4$, and $w(F) = 4$), it selects the largest initial w as its own w_r (i.e., $w_r(B) = 4$). Similarly, we have $w_r(C) = 4$, $w_r(D) = 3$, $w_r(E) = 4$, and $w_r(F) = 5$. Then, all nodes except node A use their w_r to construct the broadcasting sequences based on Algorithm 1. On the other hand, since each sender uses its calculated initial w as w_s, we have $w_s(A) = 3$, $w_s(B) = 3$, $w_s(C) = 5$, $w_s(D) = 4$, $w_s(E) = 2$, and $w_s(F) = 4$. Then, if a node needs to broadcast a message, it uses its w_s to construct the broadcasting sequence based on Algorithm 2.

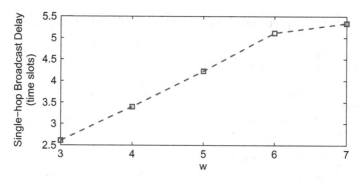

Fig. 3.3 The single-hop broadcast delay when $w_s = w_r = w$

3.1.2   The Distributed Broadcast Scheduling Scheme

Next, we consider the broadcast scheduling issue in the multi-hop broadcast scenario. The goal of the proposed distributed broadcast scheduling scheme is to intelligently select SU nodes for rebroadcasting in order to achieve the shortest broadcast delay.

First, Fig. 3.3 shows the simulation results using the parameters given in Chap. 2.5. From Fig. 3.3, we observe that the single-hop broadcast delay increases when w increases. Therefore, in a multi-hop broadcast scenario, if there are multiple intermediate nodes with the same child node, the intermediate node with the smallest w is selected to rebroadcast. If there are more than one intermediate node with the smallest w, all these nodes should rebroadcast and a broadcast collision avoidance scheme (which is explained in detail in Chap. 3.1.3) is executed before they rebroadcast the message. The pseudo-code of the proposed scheduling scheme is shown in Algorithm 3, where node v has just received the broadcast message from node q and needs to decide whether to rebroadcast. Node q includes the calculated initial w of its 1-hop neighbors in the broadcast message. Algorithm 3 indicates that each SU should know the locations of its 1-hop neighbors (in order to obtain $N(v)$) and its 2-hop neighbors (in order to obtain $N(q)$ and $d(u,k)$). Once a node receives the message, it executes Algorithm 3 to decide whether it should rebroadcast or not. If it needs to rebroadcast, it uses its calculated initial w as w_s to construct the broadcasting sequence based on Algorithm 2. Thus, as illustrated in Fig. 3.2, the message deliveries are shown by the arrows.

From the above design, it is noted that each SU (either sending or receiving) follows the same rules and no centralized entity or prior information about the sender is required. Thus, the proposed broadcast scheduling scheme is fully distributed. In addition, since the node with the smallest w is selected for rebroadcasting, the broadcast delay is the shortest. Moreover, because only a subset of intermediate nodes are selected to rebroadcast, the number of intermediate nodes that need to forward the message is reduced. Thus, the probability that multiple senders broadcasting to the same receiver simultaneously can be reduced. Hence, the proposed broadcast scheduling scheme also contributes to the broadcast collision avoidance.

Algorithm 3: The pseudo-code of the broadcast scheduling scheme for a SU sender v.

Input: $q, N(v), N(N(v)), \{w(u) | u \in N(q)\}$.
Output: Decision of rebroadcasting.

$\quad C \leftarrow \{w(v)\}$;
$\quad$ **if** $\{k | k \in (N(v) - N(v) \cap N(q))\} \neq \varnothing$ **then** /* v has at least one receiver */
$\quad\quad$ **foreach** k **do**
$\quad\quad\quad$ **if** $\{u | u \in N(q), d(u,k) \leq r_c, u \neq v\} \neq \varnothing$ **then** /* there are multiple paths from $q \rightarrow k$ */
$\quad\quad\quad\quad$ **foreach** u **do**
$\quad\quad\quad\quad\quad$ $C \leftarrow \{C, w(u)\}$;
$\quad\quad\quad\quad$ **if** $w(v) = \min C$ **and** $|\{e | e = \min C\}| = 1$ **then** /* v is the only node with the smallest w */
$\quad\quad\quad\quad\quad$ **return** TRUE;
$\quad\quad\quad\quad$ **else if** $w(v) = \min C$ **and** $|\{e | e = \min C\}| > 1$ **then** /* v is one of the multiple nodes with the same smallest w */
$\quad\quad\quad\quad\quad$ run Algorithm 4;
$\quad\quad\quad\quad\quad$ **return** TRUE;
$\quad\quad\quad\quad$ **else**
$\quad\quad\quad\quad\quad$ **return** FALSE; /* v does not rebroadcast */
$\quad\quad\quad$ **else**
$\quad\quad\quad\quad$ **return** TRUE; /* v rebroadcasts the message */
$\quad$ **else**
$\quad\quad$ **return** FALSE;

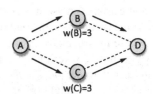

Fig. 3.4 The broadcast scenario where a broadcast collision may occur

3.1.3 The Broadcast Collision Avoidance Scheme

From Algorithm 3, if there are multiple intermediate nodes with the same child node, only the intermediate node with the smallest w should rebroadcast. However, if more than one intermediate node with the same smallest w, all these intermediate nodes should rebroadcast and a broadcast collision may occur if these nodes deliver the messages on the same channel at the same time. For instance, in the example shown in Fig. 3.4 where node A is the source node, node B and C have the same w, which may lead to a broadcast collision when they rebroadcast simultaneously.

Most broadcast collision avoidance methods in traditional ad hoc networks assign different time slots to different intermediate nodes to avoid simultaneous transmissions. However, as explained in Chap. 1, these methods cannot be applied to

CR ad hoc networks because the exact time for the intermediate nodes to receive the broadcast message is random. As a result, to assign different time slots for different intermediate nodes is very challenging. In addition, since the intermediate nodes use multiple channels for broadcasting, the channel on which the broadcast collision occurs is also unknown. To the best of our knowledge, no existing collision avoidance scheme can address these challenges in CR ad hoc networks.

In this book, we propose a broadcast collision avoidance scheme for CR ad hoc networks. The main idea is to prohibit intermediate nodes from rebroadcasting on the same channel at the same time. Our proposed broadcast collision avoidance scheme works in a scenario where the intermediate nodes have the same parent node, as shown in Fig. 3.4. The procedure of the proposed broadcast collision avoidance scheme is summarized as follows:

Step 1 Generating a Default Sequence : When a source node (e.g., node A in Fig. 3.4) broadcasts the message, it includes its own original available channel set in the message. Hence, if an intermediate node receives the message, it obtains the original available channel information of its parent node. Then, the intermediate node uses the first w available channels of its parent node to generate a default sequence, where w is its own calculated initial w (which may not be the same as the initial w of its parent node). If a channel in the default sequence is not available for this intermediate node, a void channel is assigned to replace the corresponding channel. For instance, if node B and C both obtain $w = 3$ and the original available channels of node A, B, and C are $\{1, 2, 3, 4, 5\}$, $\{2, 3, 4, 5\}$, and $\{1, 3, 4, 6\}$, respectively, node B and C only use the first three available channels of node A to generate their default sequences. Therefore, the default sequence of node B is $\{0, 2, 3\}$ and the default sequence of node C is $\{1, 0, 3\}$, where 0 means a void channel. A node does not send anything on a void channel.

Step 2 Circular Shifting the Default Sequence with a Random Number : Apart from the available channel set, the source node also includes a distinctive integer for each intermediate node v randomly selected from $[1, w(v)]$. If there are more than $w(v)$ intermediate nodes, the parent node randomly selects $w(v)$ of them and assigns a random integer. Only those intermediate nodes that acquire the random integer will rebroadcast the packet. Then, each intermediate node generates a new sequence from its default sequence using circular shift and the random integer. If we denote the default sequence as DS and the random integer as R, the intermediate node performs circular shift on the DS for R times (there is no difference of right-shift or left-shift). For instance, if node B and C get 3 and 1 as their random integers, respectively, the new sequences they generate from left-handed circular shift are $\{0, 2, 3\}$ and $\{0, 3, 1\}$, respectively.

Step 3 Forming the Broadcasting Sequence : Denote the starting time slot of the source node's broadcasting sequence as st and the time slot when an intermediate node receives the broadcast message as rt. The source node includes its st in the broadcast message. Then, the intermediate node performs circular shift on the new sequence generated from Step 2 for another $(rt - st + 1)$ times. It repeats that sequence for $w(v)$ times to form a cycle of its broadcasting sequence.

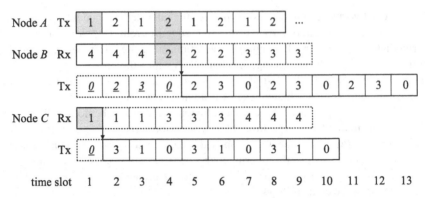

Fig. 3.5 An example of the proposed broadcast collision avoidance scheme

The pseudo-code of the broadcast collision avoidance scheme is shown in Algorithm 4, where q is the source node and Circshift() is the function of circular shift. To further elaborate the scheme, Fig. 3.5 shows an example of the proposed broadcast collision avoidance scheme. Without loss of generality, the starting time slot of the source node is 1. When node B and C do not receive the broadcast message, they hop through the channels based on the broadcasting sequences generated from Algorithm 1. Then, node B and C receive the broadcast message at time slot 4 and 1, respectively. Based on Algorithm 4 and if the random integers for node B and C are 3 and 1, respectively, node B forms the broadcasting sequence as $\{2, 3, 0, 2, 3, 0, 2, 3, 0\}$ and node C forms the broadcasting sequence as $\{3, 1, 0, 3, 1, 0, 3, 1, 0\}$. Then, they start rebroadcasting from time slot 5 and 2 using the broadcasting sequences, respectively. The underlined channels are those a node hops on if it starts from time slot 1.

Algorithm 4: The pseudo-code of the broadcast collision avoidance scheme for SU v.

Input: $q, L_q, L_v, st_q, rt_v, R_v, w(v)$.
Output: BS'_v.

$\quad BS'_v \leftarrow \varnothing;$ /* initialization */
$\quad i \leftarrow 1;$
$\quad l \leftarrow 1;$
$\quad$ **while** $i \leq w(v)$ **do**   /* generating a default sequence */
$\qquad j \leftarrow 1;$
$\qquad$ **while** $j \leq w(v)$ **do**
$\qquad\quad$ **if** $L_v(i) = L_q(j)$ **then**
$\qquad\qquad DS_v(j) \leftarrow L_q(j);$

$\quad T_v \leftarrow$ Circshift$(DS_v, R_v);$ /* circular shifting */
$\quad$ **while** $l \leq w(v)^2$ **do**   /* forming a broadcast sequence */
$\qquad BS'_v(l) \leftarrow T_v(l + (rt_v - st_q) + 1 \mod w(v));$
$\qquad l \leftarrow l + 1;$

$\quad$ **return** $BS'_v;$

Therefore, by constructing the broadcasting sequences from the same channel set (the channel set of the common parent node, node A) but circular shifting different

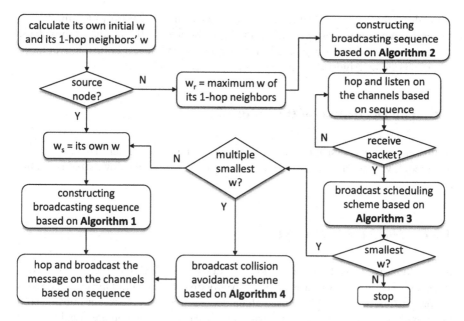

Fig. 3.6 The flow chart of the proposed BRACER protocol

times for different nodes, the intermediate nodes are guaranteed not to send on the same channel at the same time. Thus, broadcast collisions can be avoided. In addition, the proposed broadcast collision avoidance scheme still works when intermediate nodes are not synchronized. They can be synchronized based on the time stamp received from the common parent node. In this way, time slots of the intermediate nodes are perfectly aligned. Then, broadcast collisions are resolved. A trade-off of the proposed broadcast collision avoidance scheme is that less available channels are used for broadcasting because some void channels may be assigned. However, the benefit (e.g., the increase of the successful broadcast ratio) gained from eliminating broadcast collisions is greater than the loss of a very few number of channels. Hence, the only issue left is the derivation of the initial w, which is introduced in Chap. 3.2.

In this chapter, we summarize the procedure of the proposed BRACER protocol. Figure 3.6 shows the flow chart of the BRACER protocol. As shown in Fig. 3.6, before a broadcast starts, every SU node first calculates its own initial w and the initial w of its 1-hop neighboring nodes using the 2-hop location information. If this node is the source node, it uses its own initial w as its w_s and constructs the broadcasting sequence based on Algorithm 2. Then, it hops and broadcasts a message on each channel during one time slot following its sequence. On the other hand, if this node is not the source node, it is by default a receiver. Then, it uses the maximum w of its 1-hop neighboring nodes as its w_r and constructs the broadcasting sequence based on Algorithm 1. It hops and listens on each channel during one time slot following its sequence. If the node receives the broadcast message from a sender, it runs the broadcast scheduling scheme based on Algorithm 3 to determine whether it needs to rebroadcast this message. If it needs to rebroadcast and there is only one smallest w, it uses its own w as w_s and runs Algorithm 2 to rebroadcast. If it needs to rebroadcast and there are more than one smallest w, it runs the broadcast collision avoidance scheme based on Algorithm 4 to rebroadcast the message.

3.2   The Derivation of the Value of w

In this chapter, we first introduce a network model we consider. Then, based on this model, we present the derivation process of the size of the downsized available channel set w.

3.2.1   The Network Model

In this book, we consider a CR ad hoc network where N SUs and K primary users (PUs) co-exist in an $\alpha \times \alpha$ area. PUs are evenly distributed within the area. The SUs opportunistically access M licensed channels. Each SU has a circular transmission range with a radius of r_c. The SUs within the transmission range are considered as the neighboring nodes of the corresponding SU. That is, only when a SU receiver is within the transmission range of a SU transmitter, the signal-to-noise ratio (SNR) at the SU receiver is considered to be acceptable for reliable communications. In addition, apart from the broadcast collision, other factors may also contribute to the packet error (e.g., channel quality, modulation schemes, and coding rate). However, in this book, we only consider broadcast collisions as the reason for the packet error. We claim that this is a valid assumption in most broadcast scenarios [2–14].

Each SU also has a circular sensing range with a radius of r_s. That is, if a PU is currently active within the sensing range of a SU, the corresponding SU is able to detect its appearance. Since different SUs have different local sensing ranges which include different PUs, their acquired available channels may be different [15, 16]. In addition, because the available channels of a SU are obtained based on the sensing outcome within the sensing range, a SU is not allowed to communicate with other SUs outside its sensing range since it may mistakenly use an occupied channel by a PU, which results in interference to the PU. Therefore, in this book, we assume that $r_c \leq r_s$.

In this book, we model the PU activity as an ON/OFF process, where the length of the ON period is the length of a PU packet. The length of the ON period and the OFF period can follow arbitrary distributions. We assume that each PU randomly selects a channel from the spectrum band to transmit one packet which consists of multiple time slots. Moreover, because PUs at different locations can claim any channels for communications, the packets on the same channel do not necessarily belong to the same PU. This is a more practical scenario, as compared to some papers which assume that each channel is associated with a different PU. Under such a practical scenario, only those PUs that are within the sensing range of a SU and are active during the broadcast process contribute to the unavailable channels of the SU [17].

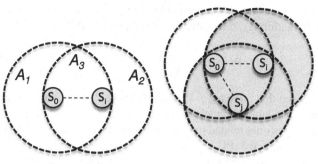

a The single-pair scenario. **b** The multi-pair scenario.

Fig. 3.7 The single-hop broadcast scenario

3.2.2 The Derivation of the Value of w

As explained in Chap. 3.1, the value of w is essential to ensure a successful single-hop broadcast. Denote the probability of a successful single-hop broadcast as $P_{succ}(w)$, where $P_{succ}(w)$ is a function of w. Our goal is to obtain an appropriate w that satisfies the condition: $P_{succ}(w) \geq 1 - \varepsilon$, where ε is a small pre-defined value. From Theorem 1, the condition that at least one common channel exists between the downsized available channel sets of a SU pair is a necessary condition for a successful single-hop broadcast. Therefore, if we denote the source SU of a single-hop broadcast as S_0 and the neighbors of S_0 as $\{S_1, S_2, \cdots, S_H\}$, where H is the number of neighbors, $P_{succ}(w)$ is equal to the probability that there is at least one common channel between S_0 and each of its neighbors in their downsized available channel sets.

We first calculate the probability that there is at least one common channel between the downsized available channel sets of S_0 and one of its neighbors S_i. The relative locations of the two SUs and their sensing ranges are shown in Fig. 3.7a. As illustrated in Fig. 3.7a, sensing ranges are divided into three areas: A_1, A_2, and A_3. Note that PUs in different areas have different impact on the channel availability of the two SUs. For instance, if a PU is active within A_3, the channel used by this PU is unavailable for both SUs. However, if a PU is active within A_1, the channel used by this PU is only unavailable for S_0. Thus, we first calculate the probability that a channel is available within each area, $P_k, k \in [1, 2, 3]$. The size of the total network area is denoted as A_L (i.e., $A_L = \alpha^2$). Since the locations of PUs are evenly distributed, the probability that p PUs are within A_k is

$$\Pr(p) = \binom{K}{p} \left(\frac{A_k}{A_L}\right)^p \left(\frac{A_L - A_k}{A_L}\right)^{K-p}, \tag{3.1}$$

where $\binom{K}{p}$ represents the total combinations of K choosing p. In addition, we define the probability that a PU is active, ρ, as:

$$\rho = \frac{E[\text{ON duration}]}{E[\text{ON duration}] + E[\text{OFF duration}]}, \tag{3.2}$$

where $E[\cdot]$ represents the expectation of the random variable. Therefore, given that there are p PUs within A_k, the probability that there are b PUs active is

$$\Pr(b|p) = \binom{p}{b}\rho^b(1-\rho)^{p-b}. \tag{3.3}$$

Furthermore, given that there are p PUs and b active PUs within A_k, the probability that there are c available channels is denoted as $\Pr(c|p,b)$. Since the number of available channels is only related to the number of active PUs, c is independent of p. In addition, since an active PU randomly selects a channel from M channels in the band, $\Pr(c|p,b)$ is equivalent to the probability that there are exactly c empty boxes given that b balls are randomly put into a total of M boxes and a box can have more than one ball (because we do not limit a channel to only one PU). Thus, $\Pr(c|p,b)$ can be expressed as:

$$\Pr(c|p,b) = \frac{\binom{M}{c}(M-c)!S(b,M-c)}{M^b}, c \in [\max(0,M-b),M], \tag{3.4}$$

where $S(b,M-c)$ is the Stirling number of the second kind. In addition, $S(b,M-c)$ is defined as

$$S(b,M-c) = \frac{1}{(M-c)!}\sum_{i=0}^{M-c}(-1)^i\binom{M-c}{i}(M-c-i)^b. \tag{3.5}$$

Hence, the probability that there are c available channels and there are p PUs and b active PUs within A_k is the product of (3.1), (3.3), and (3.4). Then, the probability that a channel is available within A_k is obtained from (3.7).

Next, we consider the relationship between the downsized available channel sets of the two SUs. In our derivation, we only consider the scenario where the sender and its receiver have the same w (i.e., $w_s = w_r$). If $w_r > w_s$, the channels after the first w_s channels do not affect the number of common channels. Thus, the derivation process is the same. Figure 3.8 shows an example of the channel availability status of two SUs when $w(S_0) = 3$, where a shaded square indicates an idle channel and a white square indicates a busy channel. A square with a cross means that a channel can be either idle or busy. Since each SU only selects the first w available channels to form a downsized available channel set, the availability status of the channels after the first w available channels is not specified. Then, without loss of generality, we denote t and h as the index of the last available channel in the downsized available channel sets of S_0 and S_i, respectively. We first assume that $t \leq h$. Hence, from channel 1 to t, there are four possible scenarios of every channel in terms of its availability for the two SUs. They are: (1) the channel is available for both SUs (denoted as $C1$); (2) the channel is unavailable for both SUs (denoted as $C2$); (3) the channel is only available for S_0 (denoted as $C3$); and (4) the channel is only available for S_i (denoted as $C4$). In addition, from channel $t+1$ to h (if $t < h$), there are two possible scenarios: (1) the channel is available for S_i but it can be any status for S_0 (denoted as $C5$) and (2) the channel is unavailable for S_i but it can be any status for S_0 (denoted as $C6$).

Fig. 3.8 An example of the channel availability status when $w(S_0) = 3$

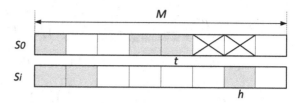

Based on Fig. 3.7a, the probabilities of the above six scenarios can be obtained: (1) $P_{C1} = P_1 P_2 P_3$; (2) $P_{C2} = (1 - P_3) + (1 - P_1)(1 - P_2)P_3$; (3) $P_{C3} = P_1 P_3 (1 - P_2)$; (4) $P_{C4} = (1 - P_1)P_2 P_3$; (5) $P_{C5} = P_{C1} + P_{C4}$; and (6) $P_{C6} = P_{C2} + P_{C3}$.

Denote $Z(0, i)$ as the number of common channels between S_0 and S_i in their downsized available channel sets. In order to obtain $\Pr(Z(0, i) = z)$, we need to consider all the combinations of the channel status for every channel from channel 1 to h. There are two possible cases: (1) $t = h$ and (2) $t < h$. For the first case, channel h is a common channel between the two SUs. In addition, from channel 1 to channel $h - 1$, there must be $z - 1$ channels in scenario $C1$; $h - 2w + z$ channels in $C2$, and $w - z$ channels in $C3$ and $C4$, respectively. Since $t = h$, no channel is in scenario $C5$ or $C6$. Thus, the probability that there are $z(z > 0)$ common channels in the first case is

$$P'(h) = \binom{h-1}{z-1}\binom{h-z}{w-z}\binom{h-w}{w-z} P_{C1}^z P_{C2}^{h-2w+z} P_{C3}^{w-z} P_{C4}^{w-z}. \tag{3.6}$$

For the second case, since $t < h$, the common available channels can only be between channel 1 to t. We denote the number of available channels for S_i from channel 1 to t as x. Thus, from channel 1 to t, similar to the first case, there are z channels in $C1$; $t - w - x + z$ channels in $C2$; $w - z$ channels in $C3$; and $x - z$ channels in $C4$. In addition, from channel $t + 1$ to h, there are $w - x$ channels in $C5$ and $h - t - w + x$ channels in $C6$. Therefore, the probability that there are totally z common channels is obtained from (7). If we switch S_0 and S_i in Fig. 3.8, we can obtain the probability for the dual case. Hence, the probability that there are z common channels in the second case is expressed in (3.9).

$$P_k = \frac{1}{M} \sum_{p=0}^{K} \sum_{b=0}^{p} \sum_{c=\max(0,M-b)}^{M} \frac{c\binom{M}{c}(M-c)!S(b, M-c)}{M^b} \binom{p}{b} \rho^b (1 - \rho)^{p-b} \tag{3.7}$$

$$\binom{K}{p}\left(\frac{A_k}{A_L}\right)^p \left(\frac{A_L - A_k}{A_L}\right)^{K-p}.$$

$$P_1''(h) = P_{C1}^z P_{C3}^{w-z} \sum_{t=w}^{h-1} \sum_{x=\max(0,w+t-h)}^{t-w} \binom{t-1}{w-1}\binom{w}{z}\binom{t-w}{x-z}\binom{h-t-1}{w-x-1} \tag{3.8}$$

$$P_{C4}^{x-z} P_{C2}^{(t-w-x+z)}(P_{C1} + P_{C4})^{(w-x)}(P_{C2} + P_{C3})^{(h-t-w+x)}.$$

$$P''(h) = P_{C1}^z P_{C3}^{w-z} \sum_{t=w}^{h-1} \sum_{x=\max(0,w+t-h)}^{t-w} \binom{t-1}{w-1}\binom{w}{z}\binom{t-w}{x-z}\binom{h-t-1}{w-x-1} \quad (3.9)$$

$$P_{C4}^{x-z} P_{C2}^{(t-w-x+z)}(P_{C1}+P_{C4})^{(w-x)}(P_{C2}+P_{C3})^{(h-t-w+x)} +$$

$$P_{C1}^z P_{C4}^{w-z} \sum_{t=w}^{h-1} \sum_{x=\max(0,w+t-h)}^{t-w} \binom{t-1}{w-1}\binom{w}{z}\binom{t-w}{x-z}\binom{h-t-1}{w-x-1} \quad (3.10)$$

$$P_{C3}^{x-z} P_{C2}^{(t-w-x+z)}(P_{C1}+P_{C3})^{(w-x)}(P_{C2}+P_{C4})^{(h-t-w+x)}.$$

Therefore, the probability that there are z common channels for the first w available channels for each SU is

$$\Pr(Z(0,i)=z) = \sum_{h=2w-z}^{M} P'(h) + P''(h). \quad (3.11)$$

Thus, the probability of a successful single-hop broadcast from S_0 to S_i is

$$P_{succ}(w) = 1 - \Pr(Z(0,i)=0). \quad (3.12)$$

Figure 3.9(a) shows the analytical and simulation results of $P_{succ}(w)$ in the single-pair scenario under various w and different M. To obtain these results, the number of PUs $K = 40$ and the probability that a PU is active $\rho = 0.9$. In addition, the side length of the network area $\alpha = 10$ (unit length) and two neighboring SUs are at the border of each other's sensing range where $r_s = 2$ (unit length). As shown in Fig. 3.9a, the simulation results match extremely well with the analytical results.

We extend the above results to a multi-pair scenario, as shown in Fig. 3.7b where S_i and S_j are two neighbors of S_0. Based on the knowledge of combination mathematics, the probability of a successful broadcast in the multi-pair scenario shown in Fig. 3.7b is

$$P_{succ}(w) = 1 - \Pr(Z(0,i)=0) - \Pr(Z(0,j)=0)$$
$$+ \Pr(Z(0,i,j)=0), \quad (3.13)$$

where $\Pr(z(0,i,j) = 0)$ is the probability that both S_i and S_j do not have any common channel in the downsized available channel sets with S_0. Since the other two terms in (3.13) (i.e., $\Pr(Z(0,i) = 0)$ and $\Pr(Z(0,j) = 0)$) can be obtained from (3.11), we only need to calculate $\Pr(Z(0,i,j) = 0)$.

To calculate $\Pr(Z(0,i,j) = 0)$, we use the same idea from the single-pair scenario. That is, we consider S_i and S_j together as one new neighboring node. The sensing range of the new neighboring node is the union of the sensing ranges of the two original nodes (i.e., the shaded area in Fig. 3.7b). Therefore, we can obtain new P_1, P_2, and P_3 for the multi-pair scenario based on the new size of the sensing range. Moreover, the probabilities of every scenario of the channel status can also be obtained accordingly. Therefore, by using (3.6–3.11), we can calculate $\Pr(Z(0,i,j) = 0)$. Then, given the locations of the H neighbors, each SU can get

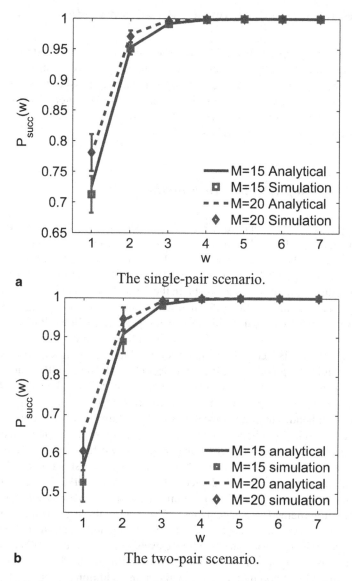

Fig. 3.9 Analytical and simulation results of $P_{succ}(w)$ under various w and different M

the probability of a successful single-hop broadcast by performing the same procedure iteratively for H times. Finally, by letting $P_{succ}(w) \geq 1 - \varepsilon$, a proper w can be acquired for S_0. Figure 3.9b shows the analytical and simulation results of $P_{succ}(w)$ in the two-pair scenario under various w and different M. From Fig. 3.9b, the simulation results match very well with the analytical results.

3.3   Discussion on the Proposed Broadcast Protocol

It is noted that our proposed BRACER protocol is particularly designed for broadcast scenarios in multi-hop CR ad hoc networks without a common control channel. As described in Chaps. 1 and 3.1, there are two implementation issues that are essential to the performance of our proposed distributed broadcast protocol: (1) the 2-hop location information; and (2) the time synchronization. In this chapter, we provide a further discussion on these two issues.

3.3.1   The 2-hop Location Information

From Chap. 3.1, in our proposed BRACER protocol, every SU node needs the location information of its 2-hop neighboring nodes in order to calculate the size of the downsized available channel sets of its 1-hop neighboring nodes. Even though the localization issue for CR ad hoc networks is out of the scope of this book, we hereby introduce several solutions to obtain the 2-hop location information in detail. Generally speaking, the location information for a traditional ad hoc network can be obtained either from external positioning techniques (e.g., Global Positioning System (GPS) [18]) or from some localization algorithms without external positioning techniques [19, 20]. Hence, GPS is an option to obtain the location information of the 2-hop neighboring nodes in CR ad hoc networks. However, GPS requires additional hardware and consumes extra energy, which may not be efficient in CR ad hoc networks where cost and power constraints are often needed.

On the other hand, a number of localization algorithms that do not rely on GPS for CR ad hoc networks have been proposed [21, 22]. In these works, the legacy localization algorithms proposed for traditional ad hoc networks, such as time-of-arrival (TOA)-based, angle-of-arrival (AOA)-based, and received-signal-strength (RSS)-based methods are improved and adopted in CR ad hoc networks. These localization algorithms often require the assistance from certain special nodes with known location information (named reference nodes). However, all these algorithms ignore the control message exchange issue between the reference nodes and the regular nodes in CR ad hoc networks. The control message exchange issue is either not considered or simplified by using a common control channel. Based on Chap. 1, transmitting messages on a global common channel without any additional control information is not feasible in CR ad hoc networks. Therefore, in order to receive the control message containing the location information from the reference nodes, a communication mechanism that does not rely on any other control information (i.e., under blind information) between the reference nodes and the regular nodes is needed. As mentioned before, in [17], a QoS-based broadcast protocol under blind information is proposed. We can use this scheme as the communication scheme between the reference nodes and the regular nodes to obtain the 2-hop location information. Since the broadcast protocol proposed in [17] can only support QoS provisioning,

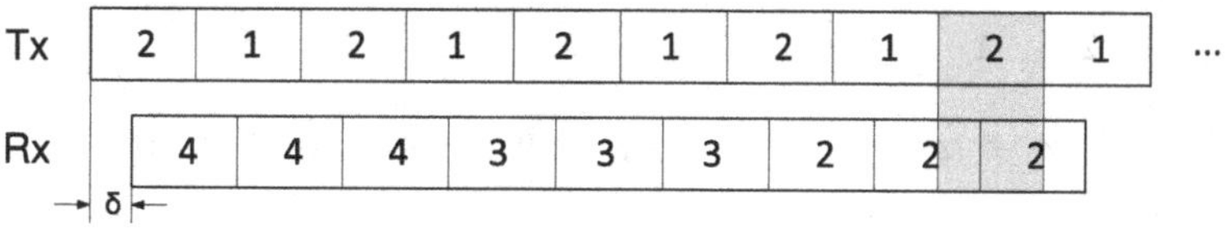

Tx	2	1	2	1	2	1	2	1	2	1	...

Rx	4	4	4	3	3	3	2	2	2

$\rightarrow$ δ $\leftarrow$

Fig. 3.10 An example of Scenario I when time slots are unsynchronized

the successful broadcast ratio and average broadcast delay of this scheme for the whole network are not optimized. Therefore, this scheme is suitable to be used in the early stage of a broadcast procedure. After every node in the network acquires the 2-hop location information, the proposed BRACER protocol can be executed.

3.3.2 Time Synchronization

From Chap. 1, an advantage of our proposed BRACER protocol is that it does not require tight time synchronization. This special advantage is essential since tight time synchronization is extremely difficult to achieve in a real ad hoc network system. In this book, we define tight time synchronization as the scenario where time slots of different nodes are precisely aligned. This means that the proposed BRACER protocol can guarantee the successful reception of a whole broadcast message even if the time slots of the sender and the receiver have an offset. Denote the length of the offset as δ. Without the loss of generality, δ is less than a time slot. Based on Theorem 1, in order to guarantee a successful single-hop broadcast, w_s must be smaller than or equal to w_r. Thus, we consider the time synchronization issue under the following two scenarios.

1) Scenario I: w_s is strictly smaller than w_r. If $w_s < w_r$ and the sender and the receiver have at least one common channel between their downsized available channel sets, we have the following theorem:

Theorem 2 *If* $w_s < w_r$, *the single-hop broadcast is a guaranteed success within* w_r^2 *time slots even if the time slots of the sender and the receiver have an offset.*

Proof Similar to the proof of Theorem 1, if $w_s < w_r$, during the w_r consecutive time slots for which the receiver stays on the same channel, every channel of the sender must appear at least once. More importantly, since δ is less than a time slot, at least a whole time slot of the common channel between the sender and the receiver must be completely covered by the w_r consecutive time slots of the common channel. That is, the receiver can hear a whole time slot of the common channel when the sender broadcasts the message. Thus, a successful single-hop broadcast is guaranteed.

Figure 3.10 shows an example of Scenario 1 where $w_s < w_r$. We assume that the time slots of the sender are ahead of the receiver with an offset of δ. As illustrated in Fig. 3.10, on the 9-th slot of the sender's broadcasting sequence, the sender and the receiver are on the same channel (i.e., channel 2). In addition, this time slot is

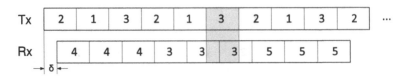

Fig. 3.11 An example of Case 1 in Scenario II when time slots are unsynchronized

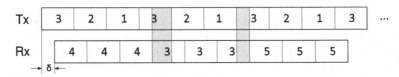

Fig. 3.12 An example of Case 2 in Scenario II when time slots are unsynchronized

completely covered by the 3 consecutive time slots when the receiver is on channel 2. Hence, the broadcast message can be successfully received by the receiver.

2) Scenario II: w_s is equal to w_r. If $w_s = w_r$, there are two sub-cases: (1) *Case 1:* a time slot of the common channel is completely covered by the w_r consecutive time slots of the receiver on the same channel; and (2) *Case 2:* a time slot of the common channel is partially covered by the w_r consecutive time slots of the receiver on the same channel. Figure 3.11 shows an example of Case 1 in Scenario II. Similar to Scenario I, the broadcast message can still be successfully received even if an offset exists.

On the other hand, Fig. 3.12 shows an example of Case 2 in Scenario II. This case occurs when the time slot of the common channel of the sender is partially covered by the first and the last time slot of the w_r consecutive time slots of the receiver. From the communication theory, if a node only receives a part of a packet, it cannot decode this packet correctly and will drop it at the physical (PHY) layer. Thus, even if the sender and the receiver have a common channel, the receiver cannot successfully receive the broadcast message within w_r^2 time slots in Case 2.

We provide two simple modifications of our proposed BRACER protocol for this case. The first way is that the receiver always shift the whole cycle of the broadcasting sequence one slot forward or one slot backward after it hops for one cycle (i.e., w_r^2 time slots) and has not received the broadcast message. At the same time, the total length of time slots that the sender broadcasts needs to be longer than three cycles of the receiver's broadcasting sequence. That is, the sender broadcasts the message following its broadcasting sequence for $\lfloor \frac{3 \times M^2}{w_s^2} \rfloor + 1$ cycles. In this way, Case 2 becomes Case 1. Then, even if the receiver may not receive the message within one cycle, it can still successfully receive the message in the following cycle, as shown in Fig. 3.13.

On the other hand, the second way is that the receiver v selects $w_r(v)$ to be $\max\{w(u)|u \in N(v)\} + 1$, where $N(v)$ is the set of the neighboring nodes of the receiver v. Therefore, the w_r of the receiver is always larger than the w_s used by the sender. In this way, Case 2 becomes Scenario I. Based on Theorem 2, the successful

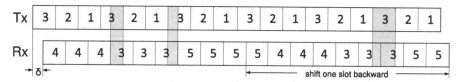

Fig. 3.13 An example of the first way of modification for Case 2 in Scenario II when time slots are unsynchronized

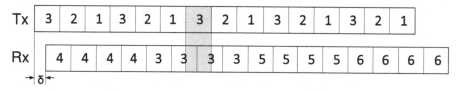

Fig. 3.14 An example of the second way of modification for Case 2 in Scenario II when time slots are unsynchronized

broadcast is guaranteed within w_r^2 time slots, as shown in Fig. 3.14. To sum up, from the above analysis, our proposed BRACER protocol can be used in an environment where tight time synchronization is not required.

3.4   Performance Evaluation

In this chapter, we evaluate the performance of the proposed broadcast protocol. We consider two types of PU traffic models in the simulation [23]. The first PU traffic model is discrete-time, where the PU packet inter-arrival time follows the biased-geometric distribution [24]. The second PU traffic model is continuous-time, where the PU packet inter-arrival time follows the Pareto distribution [24]. We assume that the probability that a PU is active is fixed (i.e., $\rho = 0.9$). In addition, the side length of the network area $\alpha = 10$ (unit length). We assume that the radius of the sensing range and the transmission range are the same (i.e., $r_s = r_c = 2$ (unit length)). In this book, we mainly investigate the following two performance metrics: (1) *successful broadcast ratio*: the probability that all nodes in a network successfully receive the broadcast message and (2) *average broadcast delay*: the average duration from the moment a broadcast starts to the moment the last node receives the broadcast message. In addition, we compare our proposed broadcast protocol with five other schemes: (1) *Random+Flooding*: each SU randomly selects a channel to hop and uses flooding (i.e., a SU is obligated to rebroadcast once receiving the message); (2) *Sequence+Flooding* (1/3 of our design): each SU downsizes its available channel set and constructs broadcasting sequences based on our scheme and uses flooding; (3) *Sequence+Schedule* (2/3 of our design): each SU constructs broadcasting sequences based on our scheme and uses our broadcast scheduling scheme; (4) *Basic QoS*

Table 3.2 Successful broadcast ratio under different number of SUs

	$N = 5$	$N = 10$	$N = 15$	$N = 20$	$N = 25$
Random+Flooding	0.8801	0.8180	0.8100	0.8726	0.8821
	0.8630	0.9148	0.9075	0.8698	0.8708
Sequence+Flooding	0.9849	0.9839	0.9828	0.9823	0.9863
	0.9762	0.9769	0.9777	0.9773	0.9719
Sequence+Schedule	0.9859	0.9864	0.9823	0.9857	0.9855
	0.9812	0.9845	0.9849	0.9876	0.9861
Basic QoS scheme	0.8915	0.9022	0.8543	0.9314	0.9317
	0.8739	0.8386	0.8952	0.8498	0.8624
Proposed scheme	0.9991	0.9973	0.9969	0.9982	0.9909
	0.9994	0.9959	0.9954	0.9967	0.9951

Scheme: each SU uses the basic scheme of the QoS-based broadcast protocol to broadcast [17]; and (5) *JS+Flooding*: each SU uses the jump-stay scheme [25] to construct the broadcasting sequences and uses flooding.

Since the single-hop successful broadcast ratio depends on w which is related to a pre-defined value ε, we define $\varepsilon = 0.001$. In fact, ε can be an arbitrary small value. Thus, based on Chap. 3.2, each SU calculates a proper w before the broadcast starts in our scheme, the *Sequence+Flooding* scheme, and the *Sequence+Schedule* scheme. Table 3.2 and 3.3 show the simulation results of the successful broadcast ratio under different number of SUs and PUs, where the value in the upper cell is for the discrete-time PU traffic and the lower cell is for the continous-time PU traffic. In Table 3.2, $M = 20$ and $K = 40$. In Table 3.3, $M = 20$ and $N = 20$. As shown in Table 3.2 and 3.3, the successful broadcast ratio is higher than 99 % under our proposed broadcast protocol in all scenarios. In addition, the proposed broadcast protocol outperforms other schemes in terms of higher successful broadcast ratio. Since the jump-stay scheme requires that the i-th available channel in the available channel set is also channel i, it cannot utilize the technique in our scheme to downsize the original available channel set. In addition, the jump-stay scheme can guarantee rendezvous within $6MP(P - G)$, where P is the smallest prime number larger than M and G is the number of common channels between two SUs. Thus, in order to ensure a successful broadcast, each SU broadcasts the message for $6MP(P - G)$ slots. However, $6MP(P - G)$ is usually a very large number when M is large. Hence, to better illustrate the trade-off between the successful broadcast ratio and broadcast delay, we compare our scheme with *JS+Flooding*.

Table 3.4 and 3.5 show the simulation results of the average broadcast delay under different number of SUs and PUs. Similarly to the successful broadcast ratio, in Table 3.4, $M = 20$ and $K = 40$. In Table 3.5, $M = 20$ and $N = 20$. As shown in Table 3.4 and 3.5, the proposed broadcast protocol outperforms other schemes in terms of shorter average broadcast delay. Furthermore, Fig. 3.15a and b show the average broadcast delay under different number of channels when $N = 10$ and $K = 40$. Besides our proposed scheme, we also compare with *JS+Flooding* and

Table 3.3 Successful broadcast ratio under different number of PUs

	$K = 20$	$K = 30$	$K = 40$	$K = 50$	$K = 60$
Random+Flooding	0.8189	0.8326	0.8842	0.9208	0.8907
	0.7980	0.8738	0.9191	0.9139	0.8849
Sequence+Flooding	0.9866	0.9863	0.9823	0.9819	0.9871
	0.9742	0.9765	0.9773	0.9711	0.9797
Sequence+Schedule	0.9868	0.9872	0.9857	0.9881	0.9872
	0.9874	0.9885	0.9876	0.9833	0.9850
Basic QoS scheme	0.9502	0.9167	0.9314	0.8222	0.7884
	0.8950	0.8921	0.8498	0.8792	0.8463
Proposed scheme	0.9978	0.9976	0.9982	0.9951	0.9921
	0.9946	0.9941	0.9967	0.9977	0.9969

Table 3.4 Average broadcast delay under different number of SUs

Delay (unit: slots)	$N = 5$	$N = 10$	$N = 15$	$N = 20$	$N = 25$
Random+Flooding	19.781	26.483	28.003	29.252	31.203
	20.981	23.765	27.686	33.153	32.883
Sequence+Flooding	8.458	11.168	12.744	14.243	15.909
	7.712	11.799	12.903	14.534	17.257
Sequence+Schedule	7.811	10.995	13.324	13.896	15.823
	7.155	11.457	13.553	14.551	15.078
Basic QoS scheme	15.576	19.642	26.447	22.745	24.599
	16.093	23.164	21.698	26.834	32.078
Proposed scheme	7.066	10.532	12.259	13.353	15.198
	6.545	11.097	12.786	13.639	14.801

our scheme without downsizing the available channel set (i.e., $w = M$). It is shown that even though the successful broadcast ratio is similar, the broadcast delay under *JS+Flooding* is much longer than our proposed scheme.

To sum up, our proposed broadcast protocol outperforms *Random+Flooding* in terms of higher successful broadcast ratio and shorter broadcast delay. It also outperforms *JS + Flooding* in terms of shorter broadcast delay. In addition, even with the trade-off in our proposed broadcast collision avoidance scheme as explained in Chap. 3.1.3 and limited overhead, our proposed scheme and the schemes that use a part of our design (e.g., *Sequence+Flooding*) can still achieve better performance results than *Random+Flooding* for both metrics and *JS+Flooding* for the broadcast delay.

From the discussion in Chap. 3.3.2, our proposed BRACER protocol has an advantage that tight time synchronization is not required. Accordingly, we provide two modifications of our proposed protocol when time slots are unsynchronized.

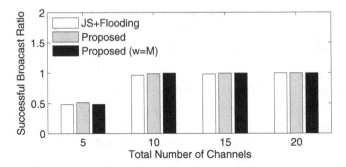

Fig. 3.15 Successful broadcast ratio under different number of channels

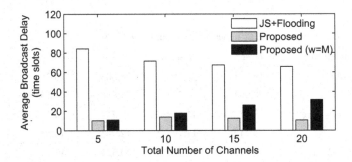

Fig. 3.16 Average broadcast delay under different number of channels

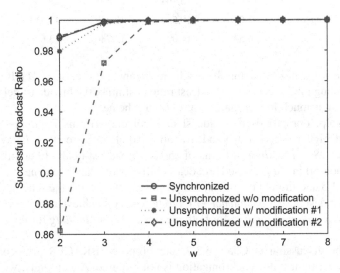

Fig. 3.17 The impact of unsynchronized time slots on the single-hop successful broadcast ratio

Table 3.5 Average broadcast delay under different number of PUs

Delay (unit: slots)	$K = 20$	$K = 30$	$K = 40$	$K = 50$	$K = 60$
Random+Flooding	29.189	31.459	25.737	25.361	24.243
	34.547	30.629	27.617	28.424	26.399
Sequence+Flooding	13.918	14.886	14.243	14.649	14.259
	14.413	13.958	14.534	14.867	14.389
Sequence+Schedule	12.747	14.206	13.896	14.361	14.014
	13.652	14.086	14.551	14.521	14.237
Basic QoS scheme	25.148	25.187	22.745	27.182	28.533
	29.111	24.931	26.834	24.639	24.907
Proposed scheme	12.322	13.555	13.352	14.279	13.597
	13.249	13.401	13.639	13.335	13.471

In this chapter, we evaluate the impact of the unsynchronized time slots on the performance of the proposed BRACER protocol.

Figures 3.17 and 3.4 show the single-hop successful broadcast ratio and the average broadcast delay under different scenarios. In the first modification, we let $w_s = w_r = w$, whereas in the second modification, we let $w_s = w$ and $w_r = w + 1$. It is shown that unsynchronized scenarios usually lead to lower successful broadcast ratio and longer average broadcast delay than the synchronized scenario. However, with the modifications of our proposed protocol, the low successful broadcast ratio can be significantly improved. From the figures, we may see that the second modification outperforms the first modification in terms of higher successful broadcast ratio. However, it also results in longer average broadcast delay than the first modification. Furthermore, when $w > 5$, the performance of the two modifications is very close to the unsynchronized scenario without modification. This is because that when w is large enough, more than one common channels exist between the sender and the receiver. Thus, there is at least one time slot on the common channel that is completely covered by the w_r consecutive time slots. Hence, the receiver can successfully receive the message without any modification.

Figures 3.18 and 3.19 show the multi-hop successful broadcast ratio and average broadcast delay under different scenarios. It is illustrated in Fig. 3.18 that when the number of SUs is small (e.g., $N < 20$), the synchronized scenario outperforms all the unsynchronized scenarios in terms of higher successful broadcast ratio. This is because when N is small, each SU usually selects small w for broadcasting. Thus, from Fig. 3.17, the successful broadcast ratio of the unsynchronized scenarios is lower than the synchronized scenario. However, when N is large (e.g., $N > 20$), the unsynchronized scenarios with both modifications outperform the synchronized scenario in terms of higher successful broadcast ratio. This is because when N is large, a receiver often has more than one senders. These senders broadcast the message on different channels to the receiver. Thus, the impact of unsynchronized time slots is diminished.

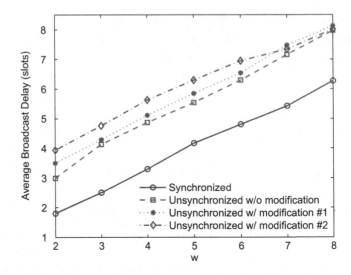

Fig. 3.18 The impact of unsynchronized time slots on the multi-hop successful broadcast ratio

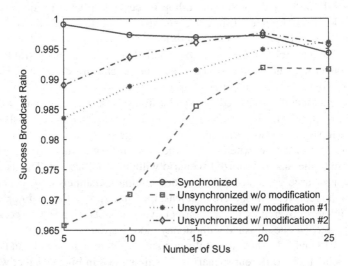

Fig. 3.19 The impact of unsynchronized time slots on the multi-hop average broadcast delay

In this chapter, we evaluate the performance of broadcast collisions for our proposed BRACER protocol. Since broadcast collisions usually lead to the waste of network resources, they should be efficiently avoided to save network resources. In this book, we use the average number of broadcast collisions in a broadcast procedure per SU node as the performance metric.

Figure 3.20 shows the average number of broadcast collisions under different numbers of channels. It is illustrated that the *Proposed Scheme* outperforms the *Sequence+Flooding* and *Sequence+Schedule* schemes in terms of fewer broadcast

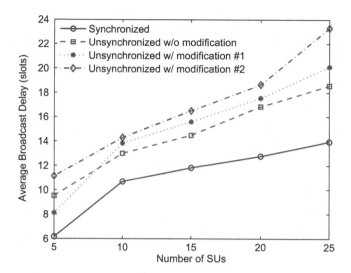

Fig. 3.20 The impact of unsynchronized time slots on the multi-hop average broadcast delay.

collisions on average. This means that the broadcast collision avoidance scheme in the *Proposed Scheme* can effectively avoid broadcast collisions. In addition, the *Proposed Scheme* also incurs fewer broadcast collisions than the *Random+Flooding* scheme when $M \leq 20$. That is, the *Random+Flooding* scheme performs better than the *Proposed Scheme* only when M is very large. This is because that in the *Random+Flooding* scheme, each sender randomly selects an available channel in the band to broadcast. If the number of channels is large, the probability that two senders select the same channel is fairly low. However, when M is small, the *Random+Flooding* scheme leads to the highest number of broadcast collisions among the four schemes (e.g., $M = 5$). Even though the *Random+Flooding* scheme causes the fewest broadcast collisions when M is large, the successful broadcast ratio and average broadcast delay of the *Random+Flooding* scheme are not acceptable, as shown in Table 3.2∼3.5. Additionally, the *Sequence+Schedule* scheme performs better than the *Sequence+Flooding* scheme, as shown in Fig. 3.20. This means that our proposed distributed broadcast scheduling scheme also contributes to the collision avoidance.

Overhead is an important metric to evaluate the efficiency of a broadcast protocol. To evaluate the impact of overhead, we use normalized overhead as the performance metric [26, 27]. Normalized overhead is defined as the ratio of the total broadcast packets (in bits) propagated by every node in the network to the total broadcast packets (in bits) received by the receivers [26, 27].

We denote the length of the original broadcast packet as L_b. Based on Chap. 3.1, extra information needs to be added in the original broadcast packet in order to realize the proposed BRACER protocol. The extra information in a broadcast packet mainly consists of three parts. First of all, as mentioned in Chap. 3.1.2, the sender should include the calculated initial w of its 1-hop neighbors in the broadcast message.

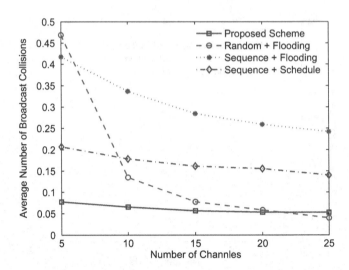

Fig. 3.21 Average number of broadcast collisions under different numbers of channels when $N = 10$

Secondly, as described in Chap. 3.1.3, the sender should include its own channel availability information and the starting time slot of its broadcasting sequence in the message. Thirdly, the sender should include random integers for the intermediate nodes who need to rebroadcast to the same node. Thus, if we define the length of the initial w, the starting time slot, and the random integer as 8 bits, the length of the total extra information in a broadcast packet in bits for a node is

$$\Theta = 8N_a + M + 8 + 8N_b, \tag{3.14}$$

where N_a is the number of the 1-hop neighbors of the node and N_b is the number of the intermediate nodes who need to rebroadcast to the same node. Therefore, the total length of a broadcast packet of the proposed BRACER protocol is $L_b + \Theta$.

Figure 3.21 shows the normalized overhead under different lengths of the original broadcast packet. We set the range of the original broadcast packet length as [50, 500] bits. Since broadcast packets are control packets which are often very short, they mainly fall in this range. In addition, we compare our proposed scheme with the *Sequence+Flooding* and *Sequence+Schedule* schemes. The *Random+Flooding* scheme does not require the 2-hop location information, so we exclude it for fair comparison. From Chap. 3.1, the length of the extra information in a broadcast packet for the *Sequence+Flooding* and *Sequence+Schedule* schemes are $\Theta = 0$ and $\Theta = 8N_a$, respectively. Thus, the *Proposed Scheme* has the longest broadcast packets among the three schemes. Even though the *Proposed Scheme* has the longest extra information in a packet, it outperforms the other two schemes in terms of lower normalized overhead, as shown in Fig. 3.21. The *Proposed Scheme* can achieve up to 106% and 12.5% less normalized overhead than the *Sequence+Flooding* and *Sequence+Schedule* schemes, respectively.

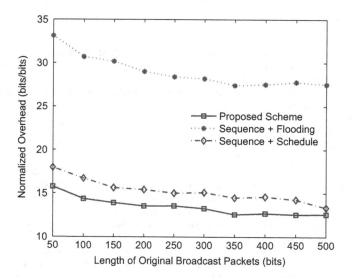

Fig. 3.22 Normalized overhead under different lengths of the original broadcast packet

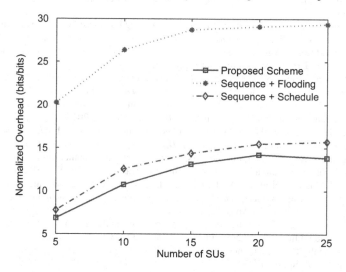

Fig. 3.23 Normalized overhead under different numbers of SUs when $L_b = 192$ bits

Figure 3.22 shows the normalized overhead under different numbers of SUs. We use the AODV route request (RREQ) packet as a typical original broadcast packet (i.e., $L_b = 192$ bits) [28]. From Fig. 3.22, it is shown that the proposed BRACER broadcast protocol outperforms the other two schemes in terms of lower normalized overhead under various numbers of SUs. More importantly, when the number of SUs increases by 400 %, the normalized overhead of the *Proposed Scheme* only increases by 115 %. Thus, the scalability of the proposed BRACER protocol is satisfactory.

References

1. Y. Song and J. Xie, "Performance analysis of spectrum handoff for cognitive radio ad hoc networks without common control channel under homogeneous primary traffic," in *Proc.* IEEE INFOCOM, 2011, pp. 3011–3019.
2. I. Chlamtac and S. Kutten, "On broadcasting in radio networks — problem analysis and protocol design," *IEEE Transactions on Communications*, vol. 33, no. 12, pp. 1240–1246, Dec. 1985.
3. S.-Y. Ni, Y.-C. Tseng, Y.-S. Chen, and J.-P. Sheu "The broadcast storm problem in a mobile ad hoc network," in *Proc.* ACM MobiCom, 1999, pp. 151–162.
4. J. Wu and F. Dai, "Broadcasting in ad hoc networks based on self-pruning," in *Proc.* IEEE INFOCOM, 2003, pp. 2240–2250.
5. J. Qadir, A. Misra, and C. T. Chou "Minimum latency broadcasting in multi-radio multi-channel multi-rate wireless meshes," in *Proc. IEEE SECON*, vol. 1, 2006, pp. 80–89.
6. A. Qayyum, L. Viennot, and A. Laouiti "Multipoint relaying for flooding broadcast messages in mobile wireless networks," in *Proc. HICSS*, 2002, pp. 3866–3875.
7. Z. Haas, J. Halpern, and L. Li "Gossip-based ad hoc routing," *IEEE/ACM Trans. Networking*, vol. 14, no. 3, pp. 479–491, June 2006.
8. Z. Chen, C. Qiao, J. Xu, and T. Lee "A constant approximation algorithm for interference aware broadcast in wireless networks," in *Proc.* IEEE INFOCOM, 2007, pp. 740–748.
9. R. Mahjourian, F. Chen, R. Tiwari, M. Thai, H. Zhai, and Y. Fang "An approximation algorithm for conflict-aware broadcast scheduling in wireless ad hoc networks," in *Proc.* ACM MobiHoc, 2008, pp. 331–340.
10. R. Gandhi, A. Mishra, and S. Parthasarathy "Minimizing broadcast latency and redundancy in ad hoc networks," *IEEE/ACM Transactions on Networking*, vol. 16, no. 4, pp. 840–851, 2008.
11. Y. Kondareddy and P. Agrawal, "Selective broadcasting in multi-hop cognitive radio networks," in *Proc.* IEEE Sarnoff Symposium, 2008, pp. 1–5.
12. C. J. L. Arachchige, S. Venkatesan, R. Chandrasekaran, and N. Mittal "Minimal time broadcasting in cognitive radio networks," in *Proc.* ICDCN, 2011, pp. 364–375.
13. Y. Song, J. Xie "A distributed broadcast protocol in multi-hop cognitive radio ad hoc networks without a common control channel," in *Proc.* IEEE INFOCOM, 2012.
14. Y. Song, J. Xie, and X. Wang "A novel unified analytical model for broadcast protocols in multi-hop cognitive radio ad hoc networks," *IEEE Transactions on Mobile Computing*, 2013.
15. Y. Song and J. Xie, "Finding out the liars: Fight against false channel information exchange attacks in cognitive radio ad hoc networks," in *Proc.* IEEE GLOBECOM, 2012, pp. 1–5.
16. C. Gao, Y. Shi, Y. T. Hou, H. D. Sherali, and H. Zhou "Multicast communications in multi-hop cognitive radio networks," *IEEE Journal on Selected Areas in Communications (JSAC)*, vol. 29, no. 4, pp. 784–793, April 2011.
17. Y. Song and J. Xie, "A QoS-based broadcast protocol for multi-hop cognitive radio ad hoc networks under blind information," in *Proc.* IEEE GLOBECOM, 2011, pp. 1–5.
18. B. H. Wellenhoff, H. Lichtenegger, and J. Collins, *Global Positions System: Theory and Practice*, 5th ed. Springer, 2001.
19. A. Savvides, C.-C. Han, and M. B. Strivastava "Dynamic fine-grained localization in ad-hoc networks of sensors," in *Proc.* ACM MobiCom, 2001, pp. 166–179.
20. T. He, C. Huang, B. M. Blum, J. A. Stankovic, and T. Abdelzaher "Range-free localization schemes for large scale sensor networks," in *Proc.* ACM MobiCom, 2003, pp. 81–95.
21. H. Celebi and H. Arslan, "Cognitive positioning systems," *IEEE Transactions on Wireless Communications*, vol. 6, no. 12, pp. 4475–4483, Dec. 2007.
22. O. Duval, A. Punchihewa, F. Gagnon, C. Despins, and V. K. Bhargava "Blind multi-sources detection and localization for cognitive radio," in *Proc.* IEEE GLOBECOM, 2008, pp. 1–5.
23. Y. Song and J. Xie, "ProSpect: A proactive spectrum handoff framework for cognitive radio ad hoc networks without common control channel," *IEEE Transactions on Mobile Computing*, vol. 11, no. 7, July 2012.
24. F. Gebali, *Analysis of Computer and Communication Networks*. Springer, 2008.

25. Z. Lin, H. Liu, X. Chu, and Y.-W. Leung "Jump-stay based channel hopping algorithm with guaranteed rendezvous for cognitive radio networks," in *Proc.* IEEE INFOCOM, 2011.
26. S.-J. Lee and M. Gerla, "Split multipath routing with maximally disjoint paths in ad hoc networks," in *Proc. IEEE ICC*, vol. 10, 2001, pp. 3201–3205.
27. C. Perkins, E. Royer, S. Das, and M. Marina "Performance comparison of two on-demand routing protocols for ad hoc networks," *IEEE Personal Communications*, vol. 8, no. 1, pp. 16–28, Feb. 2001.
28. C. E. Perkins, E. M. Belding-Royer, and S. Das "Ad hoc on-demand distance vector (AODV) routing," Request for Comments (RFC) 3561, Internet Engineering Task Force (IETF), July 2003.

Chapter 4
Unified Analytical Model for Broadcast in Cognitive Radio Ad Hoc Networks

4.1 Calculating the Successful Broadcast Ratio

In this chapter, we present the proposed algorithm for calculating the successful broadcast ratio of a broadcast protocol in multi-hop CR ad hoc networks. We first introduce a unique challenge of calculating the successful broadcast ratio. Then, the details of the proposed algorithm are presented. In addition, an example is given to show the process of the proposed algorithm. For simplicity, we assume that the wireless channels are error-free (i.e., the white noise of the channels is ignored). However, the probability that a broadcast fails due to the channel noise can be easily added in our analysis, if necessary. In the rest of the book, we use the term "sender" to indicate a SU who has just received a broadcast message and will rebroadcast the message. In addition, we use the term "receiver" to indicate a SU who has not received the broadcast message yet.

4.1.1 The Unique Challenge

Let $G(V, E)$ denote the topology of a CR ad hoc network, where V is the set of all SU nodes in the network and E is the set of all links in the network. The problem of calculating the successful broadcast ratio is described as: given a CR ad hoc network $G(V, E)$, from the source node v_s, every other node follows a certain rule to rebroadcast (e.g., simple flooding or the broadcast scheduling algorithm used in the distributed broadcast scheme in [1]), what is the successful broadcast ratio of $G(V, E)$?

As mentioned in Chap. 1, the single-hop successful broadcast ratio may not always be one in CR ad hoc networks due to various reasons. Therefore, a SU may not be able to receive the broadcast message from its direct parent node. However, during the broadcast procedure, it may receive the message from other nodes via different paths in the network. This is different from the broadcast schemes in traditional MANETs, where nodes usually receive broadcast messages from their parent nodes. This feature imposes a special challenge of calculating the successful broadcast ratio for the whole

© The Author(s) 2014   67
Y. Song, J. Xie, *Broadcast Design in Cognitive Radio Ad Hoc Networks*,
SpringerBriefs in Electrical and Computer Engineering, DOI 10.1007/978-3-319-12622-7_4

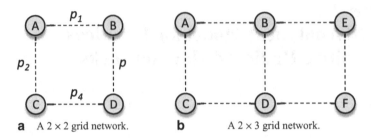

a   A 2 × 2 grid network.   **b**   A 2 × 3 grid network.

Fig. 4.1 An example for showing the unique challenge when calculating the successful broadcast ratio

CR ad hoc network. That is, there exist multiple message propagation scenarios for all the nodes to successfully receive the message. The overall successful broadcast ratio is the sum of the successful broadcast ratio of all these propagation scenarios. However, it is extremely challenging to calculate the successful broadcast ratio for every message propagation scenario when the network topology is complicated.

To further illustrate this challenge, we consider a simple 2×2 grid network shown in Fig. 4.1a, where node A is the source node. There are four links in the network, where the successful broadcast ratio over each link is given. The single-hop successful broadcast ratio depends on the specific broadcast protocol used. The method to obtain the single-hop successful broadcast ratio may be different for different protocols. We will explain the methods for calculating the single-hop successful broadcast ratio for various protocols in Chap. 4.3. If simple flooding is used to propagate the message, there are totally seven different scenarios for all nodes to successfully receive the message. They are: (1) $A \rightarrow B \rightarrow D \rightarrow C$; (2) $A \rightarrow B \rightarrow D$ and $A \rightarrow C$; (3) $A \rightarrow B$ and $A \rightarrow C \rightarrow D$; (4) $A \rightarrow C \rightarrow D \rightarrow B$; (5) $A \rightarrow B \rightarrow D$, $A \rightarrow C \rightarrow D$ and B, C do not have a collision at D; (6) $A \rightarrow C \rightarrow D \rightarrow B, A \rightarrow B$ and A, D do not have a collision at B; and (7) $A \rightarrow B \rightarrow D \rightarrow C, A \rightarrow C$ and A, D do not have a collision at C. Accordingly, since the broadcast events to different SU nodes are independent, the successful broadcast ratio for these seven scenarios is: $p_1(1 - p_2)p_3 p_4$, $p_1 p_2 p_3(1 - p_4)$, $p_1 p_2(1 - p_3)p_4$, $(1 - p_1)p_2 p_3 p_4$, $p_1 p_2 p_3 p_4 - p_{q1}$, $p_1 p_2 p_3 p_4 - p_{q2}$, and $p_1 p_2 p_3 p_4 - p_{q2}$, where p_{q1} is the probability that B and C fail to broadcast to D due to broadcast collisions and p_{q2} is the probability that A and D fail to broadcast due to broadcast collisions. The probability that two nodes have a collision also depends on the specific broadcast protocol used. Therefore, the overall successful broadcast ratio is the sum of the successful broadcast ratio of these seven scenarios, that is,

$$P_{succ} = p_1(1 - p_2)p_3 p_4 + p_1 p_2 p_3(1 - p_4) + p_1 p_2(1 - p_3)p_4 +$$
$$(1 - p_1)p_2 p_3 p_4 + (p_1 p_2 p_3 p_4 - p_{q1}) + 2(p_1 p_2 p_3 p_4 - p_{q2}). \qquad (4.1)$$

Then, we increase the dimension of the grid network to 2×3, as shown in Fig. 4.1b. If simple flooding is used, the total number of message propagation scenarios is 40. The overall successful broadcast ratio is the sum of the successful broadcast ratio of

Table 4.1 Notations used in the proposed algorithm 5

$E(v)$	The set of all the links that connect to node v
$e(v, u)$	The link that connects node v and u
$P(v, u)$	The successful broadcast ratio from node v to u
$P(G(V, E))$	The successful broadcast ratio of the network $G(V, E)$
$P_q(v, u, k)$	The probability that node v and u fail to broadcast to node k due to broadcast collisions
$\lvert \cdot \rvert$	The number of elements in a set

all these 40 message propagation scenarios. Note that although only two additional nodes and three additional links are added, the total number of propagation scenarios increases significantly. Moreover, if the grid network size is 2×4, the total number of message propagation scenarios is 252. If we further increase the dimension of the grid network to 3×3, it is almost impossible to obtain the successful broadcast ratio of every possible message propagation scenario. Therefore, when the number of nodes and links increases in a CR ad hoc network, the total number of message propagation scenarios increases exponentially. It is extremely challenging to identify every possible message propagation scenario and calculate the successful broadcast ratio for each scenario in a complicated network.

4.1.2 The Proposed Algorithm

We develop an iterative algorithm to address the above challenge. The main idea of the proposed algorithm is to decompose a complicated network into a few simpler networks so that the successful broadcast ratio of these simpler networks is straightforward to obtain and the complexity of the original network can be reduced. Then, the successful broadcast ratio of the overall network can be acquired. The notations used in the proposed algorithm are listed in Table 4.1. The pseudo-codes of the proposed algorithm for calculating the successful broadcast ratio is shown in Algorithm 5.

Under the proposed algorithm, at each iteration round, a link that connects to the source node is randomly selected. Based on whether the broadcast over this link is successful or not, the network is decomposed into two simpler networks. If the broadcast over this link is successful, all links that connect to the other node of the selected link will connect to the source node. If the broadcast over this link fails, this link is simply removed from the network. The successful broadcast ratio over each remaining link is updated accordingly after each iteration. The process terminates when only two nodes are left in the remaining networks.

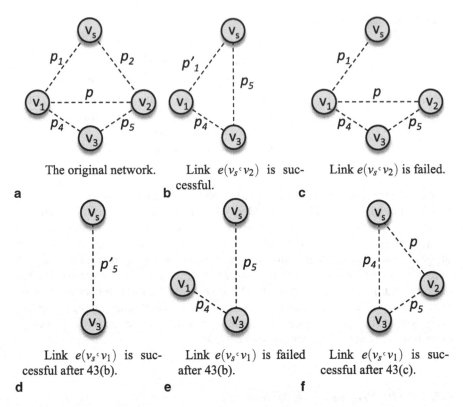

a The original network.   **b** Link $e(v_s{}^cv_2)$ is successful.   **c** Link $e(v_s{}^cv_2)$ is failed.

d Link $e(v_s{}^cv_1)$ is successful after 43(b).   **e** Link $e(v_s{}^cv_1)$ is failed after 43(b).   **f** Link $e(v_s{}^cv_1)$ is successful after 43(c).

Fig. 4.2 The process of the proposed Algorithm 5 for a 4-node CR ad hoc network

4.1.3   An Illustrative Example

We use an example to illustrate the process of the proposed Algorithm 5. As shown in Fig. 4.2a, the original CR ad hoc network consists of four nodes and five links. Based on Algorithm 5, since the source node v_s has two links, we randomly select one of these two links (e.g., link $e(v_s, v_2)$). In the first iteration, if the broadcast over the link $e(v_s, v_2)$ is successful, all nodes that are originally connected to v_2 are connected to the source node, as shown in Fig. 4.2b. In addition, the successful broadcast ratios of the new links are updated. That is, $P(v_s, v_3) = P(v_2, v_3) = p_5$ and $p'_1 = 1 - (1 - p_1)(1 - p_3) - P_q(v_s, v_2, v_1)$ because the message propagation scenarios in the original network for v_1 to successfully receive the message directly from v_s or v_2 are: (1) $v_s \rightarrow v_1$ only; (2) $v_s \rightarrow v_2 \rightarrow v_1$ only; and (3) $v_s \rightarrow v_1, v_s \rightarrow v_2 \rightarrow v_1$ and v_s, v_2 do not have a collision at v_1. The probability $(1 - p_1)(1 - p_3)$ in calculating p'_1 is the probability that both v_s and v_2 fail to broadcast to v_1. In addition, the probability that node v_s and v_2 fail to broadcast to node v_1 due to broadcast collisions $P_q(v_s, v_2, v_1)$ will be calculated in Chap. 4.3. On the other hand, if the broadcast over the link $e(v_s, v_2)$ fails, this link is simply removed from the network, as shown in Fig. 4.2c. The successful broadcast ratio of the original network can be obtained from the successful

---

Algorithm 5: The proposed algorithm for calculating the successful broadcast ratio.

Input: The topology of the network $G(V,E)$, the source node v_s.
Output: $P(G(V,E))$.
if $|V| > 2$ **then**
   if $|E(v_s)| > 1$ **then**
      $E_1 \leftarrow E; V_1 \leftarrow V;$ `/* initialization */`
      $E_2 \leftarrow E; V_2 \leftarrow V;$
      Randomly select $e(v_s, v_i) \in E(v_s)$;
      foreach $v_k, e(v_i, v_k) \in E(v_i)$ **do**
         $E_1 \leftarrow E_1 + e(v_s, v_k);$ `/* original link to` v_i `is connected to` v_s `*/`
         if $e(v_s, v_k) \in E(v_s)$ **then**
            $P(v_s, v_k) \leftarrow 1 - (1 - P(v_i, v_k))(1 - P(v_s, v_k)) - P_q(v_s, v_i, v_k);$ `/* update the link success ratio */`
         else
            $P(v_s, v_k) \leftarrow P(v_i, v_k);$
      $E_1 \leftarrow E_1 - E(v_i);$ `/* remove all links to` v_i `*/`
      $V_1 \leftarrow V_1 - v_i;$ `/* remove` v_i `*/`
      $E_2 \leftarrow E_2 - e(v_s, v_i);$ `/* remove` $e(v_s, v_i)$ `*/`
      $P(G(V,E)) \leftarrow P(v_s, v_i)P(G_1(V_1, E_1)) + (1 - P(v_s, v_i))P(G_2(V_2, E_2));$
      `/* calculate the successful ratio from the two simpler networks */`
      return $P(G(V,E))$;
   else if $|E(v_s)| = 1$ **then**
      $E_1 \leftarrow E; V_1 \leftarrow V;$
      select $e(v_s, v_i) \in E(v_s)$;
      foreach $v_k, e(v_i, v_k) \in E(v_i)$ **do**
         $E_1 \leftarrow E_1 + e(v_s, v_k);$
         $P(v_s, v_k) \leftarrow P(v_i, v_k);$
      $E_1 \leftarrow E_1 - E(v_i);$
      $V_1 \leftarrow V_1 - v_i;$
      $P(G(V,E)) \leftarrow P(v_s, v_i)P(G_1(V_1, E_1));$
      return $P(G(V,E))$;
else if $|V| = 2$ **then**
   select $e(v_s, v_i) \in E(v_s)$;
   return $P(v_s, v_i)$;   `/* iteration terminates */`

---

broadcast ratio of the two simpler networks, as shown in Fig. 4.2b and c. In the second iteration, these two simpler networks can be further decomposed following the same procedure. For the network shown in Fig. 4.2b, assume that we select the link $e(v_s, v_1)$. Similar to the process of the first iteration, this network is further decomposed into two networks, as shown in Fig. 4.2d and e, where $p_5' = 1 - (1 - p_4)(1 - p_5) - P_q(v_s, v_1, v_3)$. Then, the successful broadcast ratio of the network shown in Fig. 4.2b can be obtained from the successful broadcast ratio of these two new networks shown in Fig. 4.2d and e. For the network shown in Fig. 4.2c, since the source node has only one link, this link must be successful for other nodes to receive the message. Thus, this network is reduced to the network shown in Fig. 4.2f and the successful broadcast ratio of this

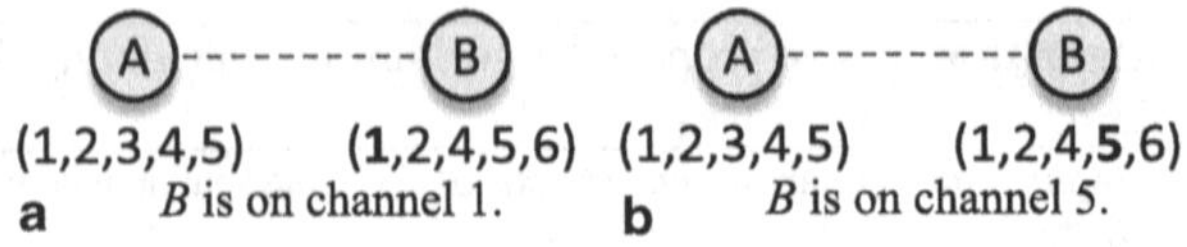

Fig. 4.3 An example for showing the randomness of the single-hop broadcast delay in CR ad hoc networks

(1,2,3,4,5)   (1,2,4,5,6)   (1,2,3,4,5)   (1,2,4,5,6)

a   *B* is on channel 1.   **b**   *B* is on channel 5.

network can be obtained from the successful ratio of the network shown in Fig. 4.2f. Therefore, if we repeat this process, the complexity of the networks from the second iteration can be further reduced. Finally, the original network can be decomposed into several single-hop networks. Then, the procedure of the proposed Algorithm 5 terminates. Therefore, the successful broadcast ratio of the original network can be expressed as

$$P_{succ} = p_2\{[1 - (1 - p_1)(1 - p_3) - P_q(v_s, v_2, v_1)][1 - (1 - p_4) \quad (4.2)$$
$$(1 - p_5) - P_q(v_s, v_1, v_3)] + [(1 - p_1)(1 - p_3) + P_q(v_s, v_2, v_1)]p_4p_5\}$$
$$+ (1 - p_2)p_1\{p_3[1 - (1 - p_4)(1 - p_5) - P_q(v_s, v_2, v_3)] + (1 - p_3)p_4p_5\}.$$

4.2 Calculating the Average Broadcast Delay

In this chapter, we introduce the proposed algorithm for calculating the average broadcast delay of a broadcast protocol. Similar to the previous chapter, we first present the unique challenge of calculating the average broadcast delay for a CR ad hoc network. Then, the detailed algorithm is given. Furthermore, an example is shown to illustrate the process of the proposed algorithm.

4.2.1 The Unique Challenge

As mentioned in Chap. 1, since the single-hop broadcast delay depends on various factors, such as the channel availability of the communication pair and specific broadcast protocol, the single-hop broadcast delay is random. Figure 4.3 illustrates the randomness of the single-hop broadcast delay in CR ad hoc networks. In Fig. 4.3, node *A* is the sender and broadcasts the message on each available channel sequentially. In addition, node *B* is the receiver and constantly listens on the channel shown in the bold font. Since node *B* does not have any information about the sender before a broadcast starts, the channel it stays on is randomly selected. It is shown that, even though the channel availability of node *B* is the same in the two scenarios shown in Fig. 4.3a and b, the single-hop broadcast delay is quite different (i.e., it takes one time slot for a successful broadcast in Fig. 4.3a, while it takes five time slots for a successful broadcast in Fig. 4.3b). Hence, due to this randomness, to obtain the single-hop broadcast delay in CR ad hoc networks is challenging. Moreover, if the number of senders and receivers is larger than one, it is even more difficult.

Fig. 4.4 An example of a
8-node CR ad hoc network
with the levels of SUs

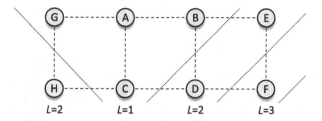

4.2.2   The Proposed Algorithm

Since to obtain the closed form expression of the average broadcast delay for arbitrary network topology is extremely complicated, in this book, we focus on the grid topology. However, the proposed methodology can be applied to any network topology. We define the level of SUs as h if they are h hops to the source node (denoted as $L = h$). Figure 4.4 shows an example of an 8-node CR ad hoc network with the levels of SUs where A is the source node. Then, the original network is decomposed into H_m levels, where H_m is the distance from the source node to the furthest node in the network. To make the derivation process tractable, we first make two assumptions. First of all, we assume that the broadcast message is propagated from the source node to the furthest node sequentially based on the relative distance to the source node. This means that, we assume that the nodes who are closer to the source node receive the message sooner than the nodes who are farther away from the source node. Based on this assumption, we categorize the SUs based on their relative distances to the source node. We further justify this assumption using simulation. We apply the broadcast protocol proposed in [2] to the network shown in Fig. 4.4. Figure 4.5 shows the simulation results of the average delay for different nodes to receive the broadcast message in the network shown in Fig. 4.4. It is shown that nodes at a higher level (e.g., nodes D and E at the second level) receive the broadcast message later than the nodes at a lower level on average (e.g., nodes B and C at the first level), which justifies our first assumption. The second assumption is that only the nodes that are at the highest level or have a path leading to the furthest node (excluding the source node) contribute to the overall average broadcast delay. Other nodes will be removed from the network for calculating the average broadcast delay. This assumption is straightforward since those nodes are independent of the message propagation path to the nodes at the highest level. For instance, in Fig. 4.4, nodes G and H do not contribute to the message propagation to node F. Thus, they can be removed when calculating the average broadcast delay of the network.

   The main idea of the proposed algorithm is that the overall average broadcast delay is the sum of the average broadcast delay at each level. At each level, it is a simple network whose average broadcast delay can be obtained. That is, $\Gamma = \sum_i^{H_m} D_i$, where Γ is the overall average broadcast delay and D_i is the average broadcast delay of the nodes at level i.

   Then, we calculate the average broadcast delay at level i, D_i. Based on the number of parent nodes, there exist only two scenarios of the single-hop broadcast in a grid

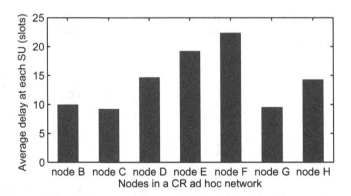

Fig. 4.5 The average delay for different nodes to receive the broadcast message in the network shown in Fig. 4.4

Fig. 4.6 Two single-hop broadcast scenarios in a grid topology network

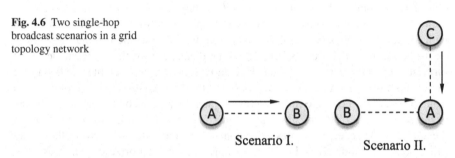

topology network. The first scenario is that a SU only has one parent node (denoted as Scenario I, as shown in Fig. 4.6a), while the second scenario is that a SU has two parent nodes (denoted as Scenario II, as shown in Fig. 4.6b). We further prove that the maximum number of parent nodes for a node in grid topology networks is two. The proof is: if there are more than two parent nodes (say, three), these three nodes should be at the same level. However, for any node that is the parent node of any two of those parent nodes (exactly 1-hop away), it needs more than two hops to reach the third parent node. That is, these three nodes cannot be at the same level. Therefore, only the two single-hop broadcast scenarios shown in Fig. 4.6 exist. We assume that for the nodes at the same level, there are α Scenario I and β Scenario II.

If the current level, level i, is not the highest level, the average broadcast delay at level i is the mean of the single-hop average broadcast delay of the nodes at level i. That is, $D_i = (\alpha \tau_1 + \beta \tau_2)/(\alpha + \beta)$, where τ_1 and τ_2 are the single-hop average broadcast delay of Scenario I and II, respectively. Denote the probabilities that the single-hop broadcast is successful at time slot k as $P_I(k)$ and $P_{II}(k)$ for Scenario I and II, respectively. $P_I(k)$ and $P_{II}(k)$ can be obtained based on a specific broadcast protocol, which is explained in Chap. 4.3. Given a successful broadcast, we first obtain the conditional probability that the single-hop broadcast is successful at time slot k for the two scenarios:

$$P_1(k) = \frac{P_I(k)}{\sum_j P_I(j)},$$

$$P_2(k) = \frac{P_{II}(k)}{\sum_j P_{II}(j)}. \tag{4.3}$$

Therefore, we have $\tau_1 = \sum_{k=1}^{T_m} k P_1(k)$ and $\tau_2 = \sum_{k=1}^{T_m} k P_2(k)$, where T_m is the maximum length of time slots the sender uses for broadcasting.

If the current level is the highest level, the calculation method for D_i is different. Since the probability that the broadcast is successful at time slot k is different in the two broadcast scenarios, we need to consider two cases: the last SU node at level i successfully receives the broadcast message is under Scenario I or Scenario II. Therefore, we first assume that the last SU node successfully receives the broadcast message at time slot d is under Scenario I and no other SU receives the message at time slot d under Scenario II. Thus, we have the probability that the single-hop broadcast delay is d at level i as

$$P'(D_i = d) = \binom{\alpha}{1} P_1(d) \left[\sum_{k=1}^{d} P_1(k) \right]^{\alpha-1} \left[\sum_{k=1}^{d-1} P_2(k) \right]^{\beta}. \tag{4.4}$$

Next, we assume that the last SU node successfully receives the broadcast message at time slot d under Scenario II and no other SU node receives the message at time slot d under Scenario I. Thus, we obtain

$$P''(D_i = d) = \binom{\beta}{1} P_2(d) \left[\sum_{k=1}^{d-1} P_1(k) \right]^{\alpha} \left[\sum_{k=1}^{d} P_2(k) \right]^{\beta-1}. \tag{4.5}$$

Last, we assume that under both scenarios, at least one node receives the broadcast message at time slot d. Hence, we have

$$P'''(D_i = d) = \binom{\alpha}{1}\binom{\beta}{1} P_1(d) P_2(d) \left[\sum_{k=1}^{d-1} P_1(k) \right]^{\alpha-1} \left[\sum_{k=1}^{d-1} P_2(k) \right]^{\beta-1}. \tag{4.6}$$

Therefore, the probability that the single-hop broadcast delay is d at level i can be written as

$$\Pr(D_i = d) = P'(D_i = d) + P''(D_i = d) + P'''(D_i = d). \tag{4.7}$$

Then, the average broadcast delay at level i is

$$D_i = \sum_{d=1}^{T_m} d \Pr(D_i = d). \tag{4.8}$$

4.2.3 An Illustrative Example

We use the example shown in Fig. 4.4 to illustrate the proposed algorithm for calculating the average broadcast delay. From Fig. 4.4, there are three levels of nodes in the network. As explained above, according to our second assumption, we first remove nodes G and H for the consideration of average broadcast delay. Then, at the first level, since both nodes B and C are under Scenario I, for D_1, we have

$$D_1 = \tau_1 = \sum_{k=1}^{T_m} \frac{k P_I(k)}{\sum_j P_I(j)}. \tag{4.9}$$

That is, the average broadcast delay at level 1 is the same as the single-hop broadcast delay under Scenario I. At the second level, nodes D and E are under different scenarios. Therefore, we have

$$D_2 = \frac{\tau_1 + \tau_2}{2} = \frac{1}{2} \left[\sum_{k=1}^{T_m} \frac{k P_I(k)}{\sum_j P_I(j)} + \sum_{k=1}^{T_m} \frac{k P_{II}(k)}{\sum_j P_{II}(j)} \right]. \tag{4.10}$$

Finally, for D_3, since this is the highest level, D_3 can be obtained using (4.8), where $\alpha = 0$ and $\beta = 1$. That is,

$$D_3 = \sum_{d=1}^{T_m} d \frac{P_{II}(d)}{\sum_j P_{II}(j)}. \tag{4.11}$$

By summing up the average broadcast delay of these three levels, the overall average broadcast delay for the network shown in Fig. 4.4 can be written as $\Gamma = \sum_{i=1}^3 D_i$.

4.3 Broadcasting in CR Ad Hoc Networks

In this chapter, we first introduce several existing broadcast designs, i.e., the random scheme and the schemes proposed in [1, 2], for CR ad hoc networks under practical scenarios. Since the broadcast schemes proposed in [3] and [4] are based on impractical assumptions (i.e., a dedicated common control channel for the whole network is employed and the available channel information of all SUs are assumed to be known), we exclude these proposals in this book. In addition, we propose the derivation methods to calculate the single-hop broadcast performance metrics (i.e., successful broadcast ratio, average broadcast delay, and broadcast collision rate) for each protocol.

4.3.1 Random Broadcast Scheme

The first broadcast scheme is called the *random broadcast scheme*. Since a SU is unaware of the channel availability information of other SUs before broadcasts are executed, a straightforward action for a SU sender is to randomly select a channel from its available channel set and broadcasts a message on that channel in a time slot. If the channel selected by the receiver is the same as the channel selected by the sender, the broadcast message can be successfully received. Figure 4.7 illustrates the procedure of the random broadcast scheme, where the shaded part represents a successful broadcast.

1) *Single-hop Successful Broadcast Ratio for the Random Broadcast Scheme:* We first calculate the single-hop successful broadcast ratio for the random broadcast scheme. Without loss of generality, in the rest of the book, the sender and the receiver of the single-hop link is denoted as A and B. We further denote the numbers of available channels for the single-hop communication pair as N_A and N_B, respectively. The number of common channels between A and B is Z_{AB}. Therefore, the probability that the single-hop broadcast is successful in a time slot is

$$p_r = \binom{Z_{AB}}{1} \frac{1}{N_A} \frac{1}{N_B} = \frac{Z_{AB}}{N_A N_B}. \tag{4.12}$$

Therefore, if the length of the time slots that the sender uses for broadcasting is S_r, the single-hop successful broadcast ratio for the random broadcast scheme is

$$P_{rand} = 1 - \left(1 - \frac{Z_{AB}}{N_A N_B}\right)^{S_r}. \tag{4.13}$$

2) *Single-hop Average Broadcast Delay for the Random Broadcast Scheme:* Next, we calculate the single-hop average broadcast delay for the random broadcast scheme. In this book, since we focus on grid topology for the broadcast delay [5], we only need to consider the two single-hop broadcast scenarios shown in Fig. 4.6. For Scenario I, since the sender and the receiver randomly select a channel in a time slot, the probability that the single-hop broadcast is successful at time slot k is $P_I(k) = (1 - p_r)^{k-1} p_r$, where p_r is given in (4.12). For scenario II, since there are two senders, we denote the other sender as C and the number of available channels of C is N_C. In addition, the number of common channels between B and C is Z_{BC}. Thus, similar to (4.12), the probability that the single-hop broadcast is successful between C and B in a time slot is $p_m = \frac{Z_{BC}}{N_B N_C}$. Hence, the probability that the single-hop broadcast is successful under Scenario II in a time slot is $p_{r2} = [1 - (1 - p_r)(1 - p_m)] - p_{q1}$, where p_{q1} is the probability that nodes A and C have a broadcast collision at node B in a time slot. The derivation of p_{q1} can also be calculated. Hence, the probability that the single-hop broadcast is successful at time slot k can be expressed as

$$P_{II}(k) = (1 - p_{r2})^{k-1} p_{r2}. \tag{4.14}$$

Then, based on (4.3), given the single-hop broadcast is successful, the conditional probability that the receiver successfully receives the broadcast message at time slot k for both scenarios under the random broadcast scheme, $P_1(k)$ and $P_2(k)$, can be obtained.

Tx	1	3	4	1	5	4	3	4	5	3	5	2
Rx	3	1	2	4	3	2	1	2	4	3	2	1

Fig. 4.7 An example of the random broadcast scheme.

3) *Single-hop Broadcast Collision Rate for the Random Broadcast Scheme:* Next, we calculate the single-hop broadcast collision rate for the random broadcast scheme. We first derive the probability that nodes A and C have a broadcast collision at node B in a time slot, p_{q1}. p_{q1} is equivalent to the probability that all the three nodes select the same channel. Denote the number of common channels among the three nodes as Z_{ABC}. Thus, we have

$$p_{q1} = \frac{Z_{ABC}}{N_A N_B N_C}. \tag{4.15}$$

Since the length of the time slots that the sender uses for broadcasting is S_r, the probability that a single-hop broadcast fails due to broadcast collisions for the random broadcast scheme can be written as

$$P_q(A, C, B) = \sum_{l=1}^{S_r} \binom{S_r}{l} p_{q1}^l \left[(1 - p_r)(1 - p_m)\right]^{S_r - l}, \tag{4.16}$$

where l is the number of time slots when nodes A and C have a broadcast collision at node B.

4.3.2 QoS-based Broadcast Scheme

The second scheme is called the *QoS-based broadcast scheme* [2, 6]. The main idea of the QoS-based broadcast scheme is to let the sender broadcast on a subset of its available channels in order to reduce the broadcast delay. In addition, the channel hopping sequences of both the sender and the receiver are designed for guaranteed rendezvous, given that the sender and the receiver have at least one channel in common in their hopping sequences. Fig. 4.7 shows an example of the QoS-based broadcast scheme. For each sender, it randomly selects n channels from its available channel set. Then, it hops and broadcasts periodically on the selected n channels for S time slots. The values of n and S are determined by the QoS requirements of the network (i.e., the successful broadcast ratio and the average broadcast delay). On the other hand, for each receiver, it first forms a random sequence that consists of its every available channel with a length of n time slots for each channel. Then, it hops and listens following this sequence periodically.

1) *Single-hop Successful Broadcast Ratio for the QoS-based Broadcast Scheme:* We continue to use the notations for calculating the single-hop performance metrics in the random broadcast scheme for the QoS-based broadcast scheme. Denote

the number of channels in the n channels selected by node A which are also in the available channel set of node B as y. We assume that the length of time slots that the sender uses for broadcasting, S, is a multiple of n. Thus, the single-hop successful broadcast ratio for the QoS-based broadcast protocol is

$$P_{qos} = \sum_{y=y^*}^{y^{**}} H(y), \qquad (4.17)$$

where $y^* = \max(1, n + Z_{AB} - N_A)$, $y^{**} = \min(n, Z_{AB})$, and $H(y)$ is written as

$$H(y) = \begin{cases} \dfrac{\binom{Z_{AB}}{y}\binom{N_A-Z_{AB}}{n-y}}{\binom{N_A}{n}} \dfrac{\binom{N_B}{y}-\binom{N_B-\frac{S}{n}}{y}}{\binom{N_B}{y}}, & \text{if } y < N_B - \dfrac{S}{n} \\[4mm] \dfrac{\binom{Z_{AB}}{y}\binom{N_A-Z_{AB}}{n-y}}{\binom{N_A}{n}}, & \text{if } y \geq N_B - \dfrac{S}{n}, \end{cases} \qquad (4.18)$$

where $\dfrac{\binom{Z_{AB}}{y}\binom{N_A-Z_{AB}}{n-y}}{\binom{N_A}{n}}$ is the probability that there are y common channels between the sender and the receiver in the selected n channels by the sender. (4.18) indicates that when S is large enough (the case when $y \geq N_B - \frac{S}{n}$), the single-hop successful broadcast ratio is independent of S.

2) *Single-hop Average Broadcast Delay for the QoS-based Broadcast Scheme:* Secondly, we calculate the single-hop average broadcast delay for the QoS-based broadcast scheme. Similar to the random broadcast scheme, we first calculate the probability that the single-hop broadcast is successful at time slot k. Based on the broadcast protocol shown in Fig. 4.7, one cycle of the broadcasting sequence of the receiver consists of N_B sections, where each section includes the same channel repeated for n times. If the channel in a section is the first appearing common available channel of nodes A and B, the single-hop broadcast is successful within that section. Denote the sections of one cycle of the broadcasting sequence of the receiver as $[f_1, f_2, \cdots, f_{N_B}]$. We calculate the probability that for a particular y, the channel in f_i is the first appearing common available channel, $\Pr(f_i), i \in [1, N_B - y + 1]$. This probability is equal to the probability that the first ball is in the i-th box if y balls are randomly put in N_B boxes. Therefore, $\Pr(f_i) = \dfrac{\binom{N_B-i}{y-1}}{\binom{N_B}{y}}$. Since time slot k is in the $(\lfloor \frac{k-1}{n} \rfloor + 1)$-th section, the probability that the single-hop broadcast is successful in $f_{\lfloor \frac{k-1}{n} \rfloor + 1}$ is $\dfrac{\binom{N_B-\lfloor \frac{k-1}{n} \rfloor - 1}{y-1}}{\binom{N_B}{y}}$. On the other hand, given that the first appearing common available channel is in $f_{\lfloor \frac{k-1}{n} \rfloor + 1}$, since the channels in the broadcasting sequence of the sender is evenly distributed, the conditional probability that the broadcast is successful in time slot k is $\frac{1}{n}$. Therefore, for Scenario I, the probability that the single-hop broadcast is successful at time slot k is expressed in (4.19).

For Scenario II, for simplicity, we assume that both the two senders have the same number of common available channels with the receiver (i.e., $Z_{AB} = Z_{BC}$). In addition, the numbers of channels that are also available for the receiver in the selected n channels by the two senders are the same (denoted as y). Denote the number of channels in the available channel sets of the two senders that are also available for all three nodes as x. Therefore, the probability that there are x channels that are available for all three nodes in their selected available channel sets is $\Pr(x) = \left(\frac{Z_{ABC}}{Z_{AB}}\right)^x \left(1 - \frac{Z_{ABC}}{Z_{AB}}\right)^{y-x}$, where Z_{ABC} is the number of channels that are available for all three nodes. Therefore, the probability that the single-hop broadcast is successful at time slot k under Scenario II is written in (4.20), where $\Pr(q)$ is the probability that there are q channels out of x channels appearing in the same time slots.

$$P_I(k) = \begin{cases} \sum_{y=y^*}^{y^{**}} \frac{\binom{Z_{AB}}{y}\binom{N_A - Z_{AB}}{n-y}}{\binom{N_A}{n}} \frac{\binom{N_B - \lfloor \frac{k-1}{n} \rfloor - 1}{y-1}}{n\binom{N_B}{y}}, & \text{if } k \leq n(N_B - y) \\ \sum_{y=y^*}^{y^{**}} \frac{\binom{Z_{AB}}{y}\binom{N_A - Z_{AB}}{n-y}}{\binom{N_A}{n}} \frac{1}{n\binom{N_B}{y}}, & \text{if } n(N_B - y) < k \leq n(N_B - y + 1) \\ 0, & \text{if } k > n(N_B - y + 1). \end{cases}$$

$$(4.19)$$

$$P_{II}(k) = \begin{cases} \sum_{y=y^*}^{y^{**}} \sum_{x=x^*}^{x^{**}} \sum_{q=0}^{q^*} \frac{\binom{Z_{AB}}{y}\binom{N_A - Z_{AB}}{n-y}}{\binom{N_A}{n}} \\ \qquad \frac{\binom{N_B - \lfloor \frac{k-1}{n} \rfloor - 1}{2y - 2q - 1}}{n\binom{N_B}{2y - 2q}} \Pr(x)\Pr(q), & \text{if } k \leq n(N_B - 2y + 2q) \\ \sum_{y=y^*}^{y^{**}} \sum_{x=x^*}^{x^{**}} \sum_{q=0}^{q^*} \frac{\binom{Z_{AB}}{y}\binom{N_A - Z_{AB}}{n-y}}{\binom{N_A}{n}} \\ \qquad \frac{1}{n\binom{N_B}{2y - 2q}} \Pr(x)\Pr(q), & \text{if } n(N_B - 2y + 2q) < k \\ & \text{and } k \leq n(N_B - 2y + 2q + 1) \\ 0, & \text{if } k > n(N_B - 2y + 2q + 1). \end{cases}$$

$$(4.20)$$

In addition, $x^* = \max(0, y - Z_{AB} + Z_{ABC})$, $x^{**} = \min(y, Z_{ABC})$, and $q^* = \min(x, y - 1)$. Thus, $\Pr(q)$ is written as

$$\Pr(q) = \begin{cases} \frac{\binom{x}{q}[(n-q)! - \sum_{j=1}^{x-q}(-1)^{(j+1)}\binom{x-q}{j}(n-q-j)!]}{n!}, & \text{if } 0 \leq q < x \\ \frac{(n-q)!}{n!}, & \text{if } q = x. \end{cases}$$

$$(4.21)$$

Then, based on (4.3), given the single-hop broadcast is successful, the conditional probability that the receiver successfully receives the broadcast message at time slot k for both scenarios under the QoS-based broadcast scheme, $P_1(k)$ and $P_2(k)$, can be obtained.

3) *Single-hop Broadcast Collision Rate for the QoS-based Broadcast Scheme:* Then, we calculate the single-hop broadcast collision rate for the QoS-based broadcast scheme. The probability that two senders have a broadcast collision is equivalent to the probability that all the common channels selected by the two senders appear in the same time slots. Therefore, using (4.21), the probability that a single-hop broadcast fails due to broadcast collisions for the QoS-based broadcast scheme is

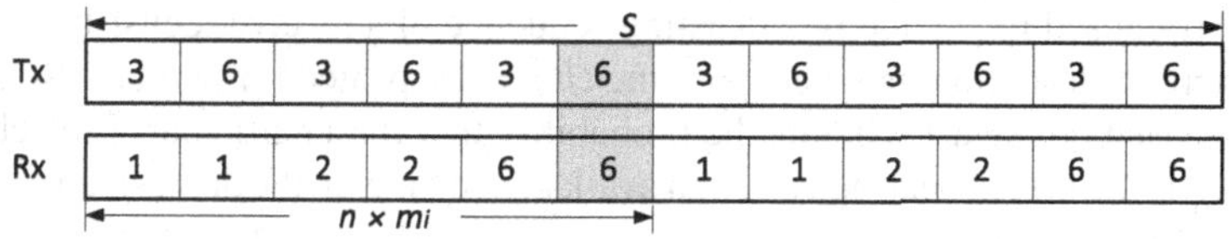

						S						
Tx	3	6	3	6	3	6	3	6	3	6	3	6
Rx	1	1	2	2	6	6	1	1	2	2	6	6

$n \times mi$

Fig. 4.8 An example of the QoS-based broadcast scheme

$$P_q(A, C, B) = \sum_{y=y*}^{y**} \frac{\binom{Z_{AB}}{y}\binom{N_A-Z_{AB}}{n-y}\binom{Z_{ABC}}{y}}{\binom{N_A}{n}\binom{Z_{AB}}{y}^2} \frac{(n-y)!}{n!}. \tag{4.22}$$

4.3.3 Distributed Broadcast Scheme

The third broadcast scheme considered in this book is called the *distributed broadcast scheme* [1, 7]. In this scheme, all SU nodes in the network intelligently select a subset of available channels from the original available channel set for broadcasting. The size of the downsized available channel set is denoted as w. The value of w needs to be carefully designed to ensure that at least one common channel exists between the downsized available channel sets of the SU sender and each of its neighboring nodes. Fig. 4.8 gives an example of the broadcasting sequences of the distributed broadcast scheme. For a SU sender, it hops periodically on the w available channels for w cycles (one cycle consists of w^2 time slots). For each receiver, it stays on one of the w available channels for w time slots. Then, it repeats for every channel in the w available channels.

1) *Single-hop Successful Broadcast Ratio for the Distributed Broadcast Scheme:* Similar to the previous schemes, we first calculate the single-hop successful broadcast ratio for the distributed broadcast scheme. As discussed above, the size of the downsized available channel set, w, has significant impact on the performance of the distributed broadcast scheme. If w is given, the single-hop successful broadcast ratio is equivalent to the probability that the sender and the receiver have at least one channel in common in their downsized available channel sets. That is, $P_{dist} = 1 - \Pr(Z(0, i) = 0)$, where $\Pr(Z(0, i) = 0)$ is the probability that the sender and the receiver do not have any common channel in their downsized available channel sets. The derivation process of $\Pr(Z(0, i) = 0)$ is the same as the method proposed in [1].

2) *Single-hop Average Broadcast Delay for the Distributed Broadcast Scheme:* Then, we calculate the single-hop average broadcast delay for the distributed broadcast scheme. For simplicity, we assume that the w obtained by the receiver is the same as the w of the sender. In addition, we denote the number of common channels between the sender and the receiver as z. We calculate the probability that the single-hop broadcast is successful at time slot k under Scenario I. Based on the broadcast

protocol proposed in [1], the broadcasting sequence of a receiver consists of w sections where each section includes the same channel repeated for w times. Similar to the QoS-based broadcast scheme, the probability that for a particular z, the channel in $t_{\lfloor \frac{k-1}{w} \rfloor + 1}$ is the first appearing common available channel in the downsized available channel set of the sender is expressed as $\Pr(t_{\lfloor \frac{k-1}{w} \rfloor + 1}) = \frac{\binom{w - \lfloor \frac{k-1}{w} \rfloor - 1}{z-1}}{\binom{w}{z}}$.

In addition, given that the first appearing common available channel is in $(\lfloor \frac{k-1}{w} \rfloor + 1)$-th section, the conditional probability that the broadcast is successful in time slot k is $\frac{1}{w}$. Therefore, for Scenario I, the probability that the single-hop broadcast is successful at time slot k is expressed as

$$P_I(k) = \begin{cases} \sum_{z=1}^{w} \frac{\binom{w - \lfloor \frac{k-1}{w} \rfloor - 1}{z-1}}{w \binom{w}{z}} \Pr(z), & \text{if } k \leq w(w-z) \\ \sum_{z=1}^{w} \frac{1}{w \binom{w}{z}} \Pr(z), & \text{if } w(w-z) < k \leq w(w-z+1) \\ 0, & \text{if } k > w(w-z+1), \end{cases} \quad (4.23)$$

where $\Pr(z)$ is the probability that there are z common channels in the downsized available channel sets between the sender and the receiver. The derivation process of $\Pr(z)$ is given in [1].

Then, for Scenario II, denote the numbers of common available channels that the two senders have with the receiver in the downsized available channel sets as z_1 and z_2, respectively. In addition, denote the number of channels in the downsized available channel sets of the two senders that are available for all three nodes as x. Since the available channels are evenly distributed in the spectrum band, the probability that there are x channels that are available for all three nodes in their downsized available channel sets is $G(x) = \binom{z^*}{x} P_A^x (1-P_A)^{z^*-x}$, where P_A is the probability that a channel is available for all three nodes and $z^* = \min(z_1, z_2)$. In addition, P_A can be obtained from [1]. Therefore, similar to the QoS-based broadcast scheme, the probability that the single-hop broadcast is successful at time slot k under Scenario II is expressed in (4.24), where $U(q)$ is the probability that there are q channels out of x channels appearing at the same time slots. In addition, $q^* = \min(x, z^* - 1)$.

$$P_{II}(k) = \begin{cases} \sum_{z_1=1}^{w} \sum_{z_2=1}^{w} \sum_{x=0}^{z^*} \sum_{q=0}^{q^*} \frac{\binom{w - \lfloor \frac{k-1}{w} \rfloor - 1}{z_1 + z_2 - 2q - 1}}{w \binom{w}{z_1 + z_2 - 2q}} \\ \quad \Pr(z_1)\Pr(z_2)G(x)U(q), & \text{if } k \leq w(w - z_1 + z_2 + 2q) \\ \sum_{z_1=1}^{w} \sum_{z_2=1}^{w} \sum_{x=0}^{z^*} \sum_{q=0}^{q^*} \frac{1}{w \binom{w}{z_1 + z_2 - 2q}} \\ \quad \Pr(z_1)\Pr(z_2)G(x)U(q), & \text{if } w(w - z_1 + z_2 + 2q) < k \\ & \text{and } k \leq w(w - z_1 + z_2 + 2q + 1) \\ 0, & \text{if } k > w(w - z_1 + z_2 + 2q + 1). \end{cases}$$
$$(4.24)$$

Using (4.21), $U(q)$ can be written as

$$U(q) = \begin{cases} \frac{\binom{x}{q}[(w-q)! - \sum_{j=1}^{x-q}(-1)^{(j+1)}\binom{x-q}{j}(w-q-j)!]}{w!}, & \text{if } 0 \leq q < x \\ \frac{(w-q)!}{w!}, & \text{if } q = x. \end{cases} \quad (4.25)$$

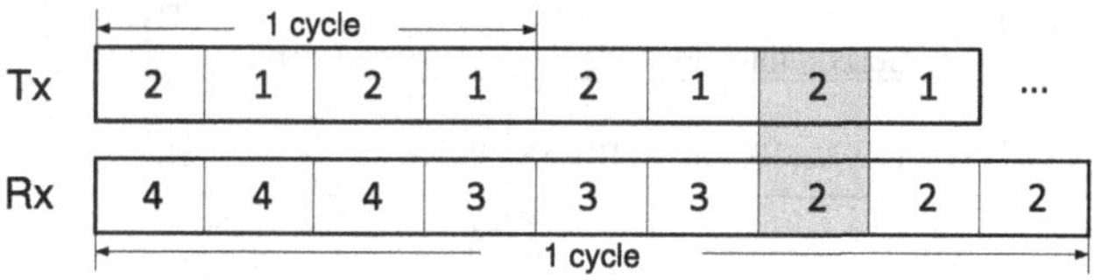

Fig. 4.9 An example of the broadcasting sequences of the distributed broadcast scheme

Then, based on (4.3), given the single-hop broadcast is successful, the conditional probability that the receiver successfully receives the broadcast message at time slot k for both scenarios under the distributed broadcast scheme, $P_1(k)$ and $P_2(k)$, can be obtained.

3) *Single-hop Broadcast Collision Rate for the Distributed Broadcast Scheme:* Finally, we calculate the single-hop broadcast collision rate for the distributed broadcast scheme. Note that in [1], a broadcast collision avoidance scheme is proposed. If this scheme is used, broadcast collisions can be avoided. However, it involves significant changes to the broadcasting sequences of the senders shown in Fig. 4.8. To make the analysis tractable, in this book, we do not consider the broadcast collision avoidance scheme. Therefore, similar to the QoS-based broadcast scheme, the probability that a single-hop broadcast fails due to broadcast collisions for the distributed broadcast scheme is

$$P_q(A, C, B) = \sum_{z=1}^{w} \frac{(w - z)!}{w!} P_A^z \Pr(z). \tag{4.26}$$

4.4 Performance Evaluation

In this chapter, we validate our proposed unified analytical model using both hardware implementation and simulation in order to prove its correctness.

4.4.1 Validating Analysis Using Hardware Implementation

The considered broadcast schemes have been implemented in embedded wireless radios. Each radio contains a Qualcomm Atheros IEEE 802.11 a/b/g chipset, and MADWIFI is used as the medium access control (MAC) driver. The three broadcast schemes are implemented as sub-functions of the MAC driver.

1) *Time Slot and Synchronization:* To support synchronized transmission of broadcast messages in different time slots, we first need to implement timing events that are synchronized among all communication nodes [8, 9]. In order to minimize the impact by the software in the driver, a hardware register called software beacon alert (SWBA) is utilized to generate timing events. To support different timing events, the

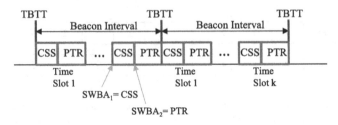

Fig. 4.10 Synchronized time slots for IEEE 802.11 chipsets

value in the SWBA register must be set into the time interval between the current timing event and the next expected timing event. Based on this mechanism, the timeline of each communication node is split into consecutive time slots each consisting of two portions: channel switching (CSS) and packet transmission/reception (PTR), as shown in Fig. 4.9.

To synchronize time slots among all nodes, we adopt two mechanisms of IEEE 802.11 [10]: target beacon transmission time (TBTT) and timing synchronization function (TSF). Within each beacon interval, the first time slot must be aligned with TBTT, as shown in Fig. 4.9. Through TSF, the time in the TSF register of different nodes is synchronized. Since TBTT is determined based on the timing value of the TSF register, the time slots of different nodes are synchronized accordingly.

2) *Packet Transmission/Reception and Channel Selection:* In a source node, a broadcast message is generated in the PTR portion of a time slot and is then sent in a selected channel. This process repeats for S time slots. Other nodes in the network attempt to receive the broadcast message from its neighboring nodes and then rebroadcast it. Due to slot-by-slot operation, when a broadcast message is received, it is rebroadcast in the next time slot in the selected channel. This process is also repeated for S time slots. Since the same message may be received for multiple times, a sequence number is added into each broadcast message to avoid redundant broadcast messages. It should be noted that the channel selection for packet transmission and reception follows the rules set by the specific broadcast schemes developed in this book. The channel set in each node reflects the activities of primary nodes and is determined according to off-line simulations.

3) *Performance Measurement:* Two performance metrics are used in our implementation: the successful broadcast ratio and the average broadcast delay. The former metric measures the probability that a broadcast message can be successfully received by all nodes in a network, and the latter one records the average delivery time from the source node to the last node. In order to get stable performance results, we repeat the experiments for N measurements as shown in Fig. 4.10. Within t_e seconds, one round of experiment is conducted. t_e is selected large enough so that all non-source nodes finish the process of receiving/rebroadcasting messages within the same period. In our experiments, we set t_e to be 3 seconds for a multi-hop CR ad hoc network under Topology 1 as shown in Fig. 4.11a.

Figure 4.12 shows comparisons between analytical results and experimental measurements for the random and QoS-based broadcast schemes. The comparisons for

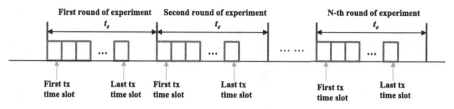

Fig. 4.11 Repeating experiments

Fig. 4.12 Topology 1 and 2
considered in the
performance evaluation

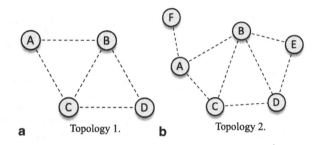

a   Topology 1.   b   Topology 2.

the distributed broadcast scheme are depicted in Fig. 4.13, where two cases are considered: (1) Case 1: all nodes have the same w (i.e., $w(A) = w(B) = w(C) =
w(D) = 5$) and (2) Case 2: some nodes have different w (i.e., $w(A) = w(B) =
w(D) = 5$ and $w(C) = 4$). As we can see from Figs. 4.12 and 4.13, the
implementation results fit the analytical results fairly well.

4.4.2   Validating Analysis Using Simulation

Due to the constraint on the total number of channels for hardware testing, we also use
simulations to validate our proposed analytical model when the number of channels
varies from 10 to 40. The side length of the simulation area $L_s = 10$ (unit length).
PUs are evenly distributed within this area. The total number of PUs is denoted as
$K = 40$. The total number of channels is denoted as M. Furthermore, each SU has a
circular transmission range with a radius of r_c. The SUs within the transmission range
are considered as the neighboring nodes of the corresponding SU. In addition, each
SU also has a circular sensing range with a radius of r_s. That is, if a PU is currently
active within the sensing range of a SU, the corresponding SU is able to detect its
appearance. Moreover, we consider the PU traffic model used in [11], where the
PU packet inter-arrival time follows the biased-geometric distribution [12, 13]. In
fact, our proposed algorithms do not rely on specific PU traffic models. We assume
that the probability that a PU is active is fixed (i.e., $\rho = 0.9$). Each PU randomly
selects a channel from the spectrum band to transmit one packet. Since the available
channels for each SU depends on the sensing outcome in its sensing range, we use the
values from the simulation as the input for the proposed analytical model (e.g., the

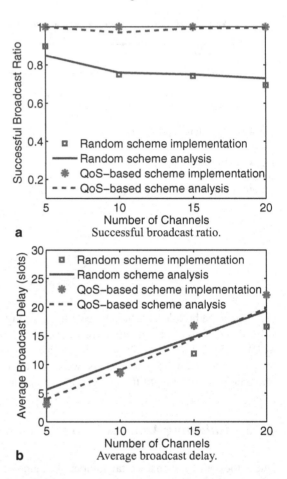

Fig. 4.13 Analytical and implementation results using the random and QoS-based broadcast schemes under Topology 1

number of common available channels between nodes A and B, Z_{AB}). In addition, we assume that the SU channel availability is stable during a broadcast duration.

1) *Single-hop Performance* We first investigate the single-hop performance of each broadcast protocol considered in this book, because this performance is the foundation of the multi-hop performance evaluation. We study the two single-hop broadcast scenarios shown in Fig. 4.6. In our study, the nodes are at the border of each other's sensing range. Fig. 4.14a , b, c show the analytical and simulation results of the single-hop successful broadcast ratio using the three considered broadcast schemes under Scenario I and II. For the random broadcast scheme, S_r is set to be the same as the number of channels, M. For the QoS-based broadcast scheme, $n = 2$ and $S = 2M$. In addition, for the distributed scheme, $w = 5$. It is shown that the simulation and analytical results match very well with the maximum difference of 0.4%, 0.5%, and 0.7% for the three schemes, respectively. The figure indicates that the distributed broadcast scheme can achieve the highest single-hop successful broadcast ratio.

Fig. 4.14 Analytical and implementation results using the distributed broadcast scheme under Topology 1

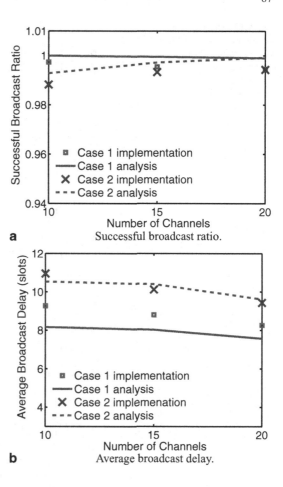

a Successful broadcast ratio.

b Average broadcast delay.

In addition, Fig. 4.15a, b, c illustrate the analytical and simulation results of the single-hop average broadcast delay using the three considered broadcast schemes under Scenario I and II. It is also shown that the simulation and analytical results match very well with the maximum difference of 1.4, 3.7, and 5.5 % for the three schemes, respectively. The distributed broadcast scheme results in the lowest single-hop average broadcast delay among the three schemes.

2) Successful Broadcast Ratio of Multi-hop CR ad hoc Networks Next, we investigate the multi-hop performance. For the successful broadcast ratio, we study the two topologies shown in Fig. 4.11a and b. The coordinates of nodes in Topology 1 are $A(4, 4)$, $B(6, 4)$, $C(5, 2.28)$, and $D(7, 2.28)$. On the other hand, note that Topology 2 is a 6-node network under arbitrary topology. Moreover, the coordinates of nodes in Topology 2 are $A(4, 4)$, $B(5.8, 4.8)$, $C(5, 3)$, $D(6.6, 3)$, $E(7, 4.5)$, and $F(3, 5)$. The parameters of each broadcast scheme are set to be the same as in the single-hop performance evaluation. In all topologies considered in the performance evaluation, node A is the source node. Fig. 4.16a , b , c show the analytical and simulation results

Fig. 4.15 Analytical and
simulation results of the
single-hop successful
broadcast ratio using the three
broadcast schemes under
Scenario I and II

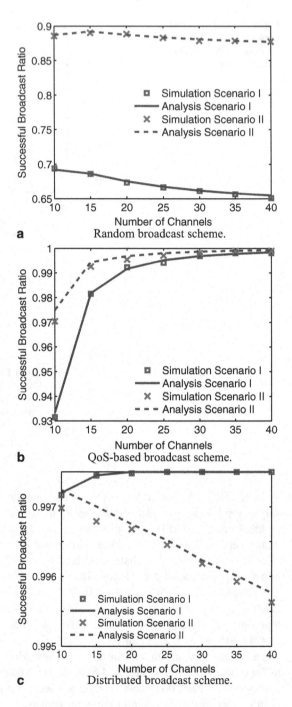

a Random broadcast scheme.

b QoS-based broadcast scheme.

c Distributed broadcast scheme.

Fig. 4.16 Analytical and simulation results of the single-hop average broadcast delay using the three broadcast schemes under Scenario I and II

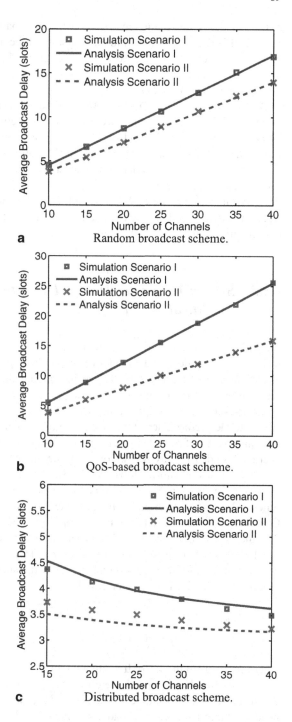

a Random broadcast scheme.

b QoS-based broadcast scheme.

c Distributed broadcast scheme.

of the broadcast ratio using the three considered broadcast schemes under Topology 1 and 2. It is shown that the simulation results fit the analytical results well with the maximum difference of 2.1, 4.6, and 0.4 % for the three schemes, respectively. The distributed broadcast scheme still has the best performance of successful broadcast ratio among the three schemes.

3) *Average Broadcast Delay of Multi-hop CR ad hoc Networks* For the average broadcast delay, we investigate two grid topology networks: (1) a 3×3 grid network (denoted as Topology 3); and (2) a 4×4 grid network (denoted as Topology 4). Figure 4.17a, b, c depict the analytical and simulation results of the average broadcast delay using the three considered broadcast schemes under Topology 3 and 4. It is shown that the simulation and analytical results coincide with each other well with the maximum difference of 4.9, 9.4, and 6.5 % for the three schemes, respectively. Again, the distributed broadcast scheme has a much lower average broadcast delay, as compared to the other two schemes.

4.4.3 System Parameter Design Using the Proposed Analytical Model

As explained in Chap. 1, the system parameters of the proposed broadcast protocols in [1–4] are not designed to achieve the optimal performance due to the lack of analytical analysis. In this book, we investigate the system parameter design of the random broadcast scheme using the proposed analytical model. In the random broadcast scheme, the length of time slots that the sender uses for broadcasting, S_r, is crucial to the performance of the broadcasting. Note that there exists a trade-off when determining S_r. If S_r is large, the successful broadcast ratio is high. However, the average broadcast delay is also long. On the other hand, if S_r is small, the average broadcast delay is short. However, the successful broadcast ratio is low. Hence, to design an optimal S_r is essential to the performance of the random broadcast scheme. We use an example to illustrate the process of the system parameter design. Consider a CR ad hoc network under Topology 1 shown in Fig. 4.11a. We assume that the single-hop successful broadcast ratio over each link is the same, which can be obtained from (4.13) (denoted as p). Thus, using the proposed algorithm for calculating the successful broadcast ratio, the successful broadcast ratio for the random broadcast scheme under Topology 1 is

$$P_{succ} = p[1 - (1 - p)^2 - P_q]^2 + p^3\{1 - [1 - (1 - p)^2 - P_q]\} \qquad (4.27)$$
$$+ (1 - p)p^2[1 - (1 - p)^2 - P_q] + (1 - p)^2 p^3,$$

where P_q is given in (4.16). It is known that P_{succ} is a function of S_r.

On the other hand, we calculate the average broadcast delay under Topology 1, where node A is the source node. Since there are two levels in the network, we need to obtain the average broadcast delay of each level. Thus, using the proposed

Fig. 4.17 Analytical and
simulation results of the
successful broadcast ratio
using the three broadcast
schemes under Topology 1
and 2

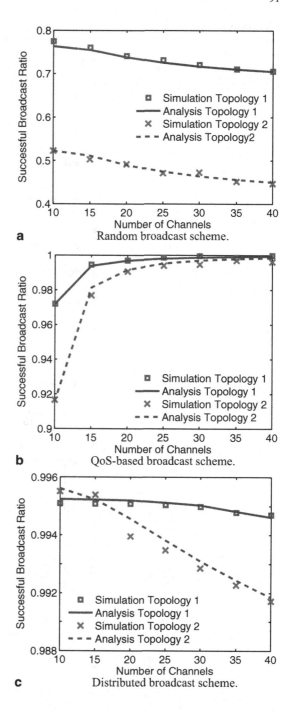

a Random broadcast scheme.

b QoS-based broadcast scheme.

c Distributed broadcast scheme.

Fig. 4.18 Analytical and
simulation results of the
average broadcast delay using
the three broadcast schemes
under Topology 3 and 4

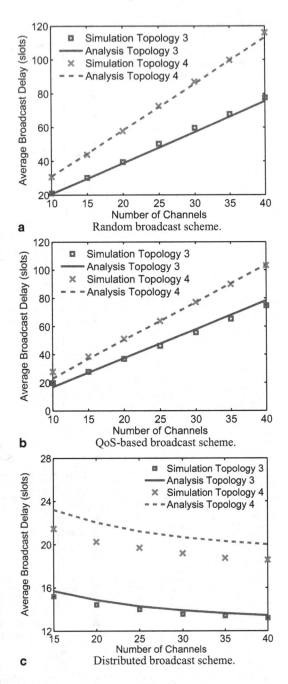

a Random broadcast scheme.

b QoS-based broadcast scheme.

c Distributed broadcast scheme.

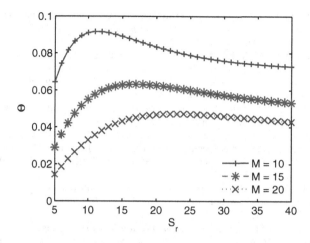

Fig. 4.19 The numerical results of the objective function under various S_r

algorithm for calculating the average broadcast delay, we have

$$\Gamma = \sum_{d=1}^{S_r} d P_1(d) + \sum_{d=1}^{S_r} d P_2(d), \tag{4.28}$$

where $P_1(d)$ and $P_2(d)$ can be obtained from (4.3). Note that Γ is also a function of S_r. Define the objective function of a broadcast protocol, Θ, as the rate between the successful broadcast ratio and the average broadcast delay. Therefore, we have $\Theta = \frac{P_{succ}}{\Gamma}$. Thus, the optimization problem of the protocol design becomes finding the optimal S_r that maximizes the objective function, Θ. Then, using certain numerical method, the optimal S_r can be obtained. Fig. 4.19 shows the numerical results of the objective function under various S_r. It is shown that a proper S_r exists to achieve the optimal performance of a broadcast protocol. For instance, when $M = 10$, the optimal S_r is 11. The corresponding successful broadcast ratio is 81.25 % and the average broadcast delay is 8.85 time slots.

References

1. Y. Song and J. Xie, "A distributed broadcast protocol in multi-hop cognitive radio ad hoc networks without a common control channel," in *Proc.* IEEE INFOCOM, 2012.
2. Y. Song and J. Xie, "A QoS-based broadcast protocol for multi-hop cognitive radio ad hoc networks under blind information," in *Proc.* IEEE GLOBECOM, 2011, pp. 1–5.
3. Y. Kondareddy and P. Agrawal, "Selective broadcasting in multi-hop cognitive radio networks," in *Proc.* IEEE Sarnoff Symposium, 2008, pp. 1–5.
4. C. J. L. Arachchige, S. Venkatesan, R. Chandrasekaran, and N. Mittal "Minimal time broadcasting in cognitive radio networks," in *Proc.* ICDCN, 2011, pp. 364–375.
5. Y. Song and J. Xie, "Optimal power control for concurrent transmissions of location-aware mobile cognitive radio ad hoc networks," in *Proc.* IEEE GLOBECOM, 2009, pp. 1–6.

6. Y. Song and J. Xie, "QB^{2}IC: A QoS-based broadcast protocol under blind information for multi-hop cognitive radio ad hoc networks," *IEEE Transactions on Vehicular Technology*, vol. 63, no. 3, pp. 1453–1466, March 2014.

7. Y. Song and J. Xie, "BRACER: A distributed broadcast protocol in multi-hop cognitive radio ad hoc networks with collision avoidance," *IEEE Transactions on Mobile Computing*, 2014.

8. X. Wang "Power efficient time-controlled CSMA/CA MAC protocol for lunar surface networks," in *Proc*. IEEE GLOBECOM, 2010, pp. 1–5.

9. Y. Song and J. Xie, "Proactive spectrum handoff in cognitive radio ad hoc networks based on common hopping coordination," in *Proc*. IEEE INFOCOM Workshops, 2010, pp. 1–2.

10. "IEEE Standard for Information Technology - Telecommunications and Information Exchange Between Systems - LAN/MAN Specific Requirements - Part 11: Wireless LAN Medium Access Control (MAC) and Physical Layer (PHY) Specifications," IEEE Std 802.11-2012 (Revision of IEEE Std 802.11-2007), 2012.

11. Y. Song and J. Xie, "ProSpect: A proactive spectrum handoff framework for cognitive radio ad hoc networks without common control channel," *IEEE Transactions on Mobile Computing*, vol. 11, no. 7, July 2012.

12. F. Gebali, *Analysis of Computer and Communication Networks*. Springer, 2008.

13. Y. Song and J. Xie, "Common hopping based proactive spectrum handoff in cognitive radio ad hoc networks," in *Proc*. IEEE GLOBECOM, 2010, pp. 1–5.

Chapter 5
Conclusion

In this book, the broadcast issue in CR ad hoc networks is investigated. Two broadcast protocols have been proposed to achieve very high successful broadcast ratio and short average broadcast delay. First of all, a QoS-based broadcast protocol named QB^2IC is proposed under practical scenarios. Moreover, a fully-distributed broadcast protocol named BRACER is proposed without the existence of a global or local common control channel. By intelligently downsizing the original available channel set and designing the broadcasting sequences and broadcast scheduling schemes, our proposed broadcast protocol can provide very high successful broadcast ratio while achieving very short broadcast delay. In addition, it can also avoid broadcast collisions. Simulation results show that our proposed protocols outperform other possible broadcast schemes in terms of higher successful broadcast ratio and shorter average broadcast delay. Finally, the performance analysis of broadcast protocols for CR ad hoc networks is studied. A novel unified analytical model is proposed to address these challenges and analyze the broadcast protocols in CR ad hoc networks with any topology. Specifically, two algorithms are proposed to calculate the successful broadcast ratio and the average broadcast delay of a broadcast protocol. In addition, the derivation methods of the single-hop performance metrics for three different broadcast protocols in CR ad hoc networks under practical scenarios are proposed. Results from both the hardware implementation and software simulation validate the analysis well. Due to the importance of broadcast operations in wireless networks, the research findings in this book will have a significant impact in future cognitive radio network designs.

© The Author(s) 2014
Y. Song, J. Xie, *Broadcast Design in Cognitive Radio Ad Hoc Networks,*
SpringerBriefs in Electrical and Computer Engineering, DOI 10.1007/978-3-319-12622-7_5